WRAP IT UP

CREATIVE STRUCTURAL PACKAGING DESIGN

 HOAKI

EDITED BY WANG SHAOQIANG

Hoaki Books, S.L.
C/ Ausiàs March, 128
08013 Barcelona, Spain
T. 0034 935 952 283
F. 0034 932 654 883
info@hoaki.com
www.hoaki.com

hoaki_books

Wrap It Up
Creative Structural Packaging Design

ISBN: 978-84-17656-31-7

Sponsored by Design 360° — Concept & Design Magazine
Edited, produced, book design, concepts & art direction by
Sandu Publishing Co., Ltd.
info@sandupublishing.com
Chief Editor: Wang Shaoqiang
Executive Editor: July Wu
Designer: Wu Yanting
Cover design by Wu Yanting
Packaging projects on the covers by Jane Wu (front cover),
SANDU Design (back cover)

D.L.: B 9920-2021
Printed in China

CONTENTS

Packaging with
Innovative Materials

Packaging with
Impressive Visuals

Die-Cut Patterns

PREFACE
PACKAGING
IS DEAD

This preface aims to work as a thinking exercise. It is interesting to create a space of reflection instead of giving comfortable and empty affirmations because, above all, first, we are thinkers, then designers. We are in a challenging moment as humanity, so we need to do the exercise of deconstructing ourselves. And as designers, we have to learn not to cling to beliefs and always continue searching for new questions, new solutions rather than fixed answers.

The purpose of the 'PACKAGING IS DEAD' statement is to provoke an internal and collective debate about our duty as designers in this uncertain and chaotic times. To start debating, we can ask: why is the packaging dead? And, if it is over, what would replace it? Should it be replaced? Is the death of the packaging a possible solution for a better world? And if it dies, what would be the containers of the products?

First of all, we have to say that death's emptiness should not be filled with anything random. We believe that we need to think more before designing because the structure we are living in is broken. When we

talk about structure, we are talking about the frantic 'consume-discard' behaviour. It is regrettable to see the garbage we are doing; for example, we reached the absurdity of creating extra packaging for fruits and vegetables. It is imperative to honestly and consciously discuss the responsibility as designers and as humans beings what we are doing with consumption and traceability. Imagine that we could reset the system and start living without packages— would it be better? It is a challenging idea, but we like it because it opens endless opportunities to rethink the industry. Packaging design is not a simple task because it has a significant impact on our life and our planet. Nowadays, there is a trend to stop that environmental impact, for example, buying at refill stations to generate fewer impact. But is it fair to recover these ancient practices? Isn't out there any other idea that can exceed this? Are the places for shopping the problem? We are convinced that we need to rethink everything. We incorporate in our daily life as normality the mechanism of buying in supermarkets physically or online, so the question is: is this the only way we can do it? Are we staying in the comfort zone? If this is something that we've created, it can be undone. We have to unlearn and start thinking if we are here to solve a problem and make it beautiful or make it better. Design in every aspect needs to think ahead and lead the change; that is the most crucial task today. It is imperative to have a holistic approach and involve other disciplines because it will give us that broad vision that we are needing.

When we make this statement of 'PACKAGE IS DEAD', you can take it literally and say that we don't need more packs in this saturated world. Or, we can use it non-literally, where we talk about taking out all our prejudice and unlearn what they taught us to transcend and innovate. By doing this, we can come to this debate with an empty mind of how packaging should look like. Sometimes our own structure and knowledge limit us, so why don't we start from scratch. We are trying to tear down our 'absolute thinking' about design. We try to apply this concept daily, so we use it to innovate if we can't use it in sustainability. In one of makebardo's project called 'Kapsi', we designed a label that does not follow the container's shape. As a strategy, we chose a standard bottle to help Kapsi fit within the sauces category, but at the same time, we decided that the label should not follow the shape of the container to stand out from the rest breaking all the design senses. This was just the beginning, the tip of the iceberg. Imagine what we could do if we went deeper? This project's result was weird and fabulous, the perfect balance to be memorable and win a prize in New Zealand. But the real question is, do we need to be memorable or more functional? We always have to be a critic of our work to evolve and understand where are we heading. With our packaging designs, we can destroy or enhance a product or a community, so we have to be careful in every step.

Working and living in different countries (Argentina, Spain, Australia and New Zealand) with varied cultures forced us to be flexible and reflective of our work as designers and our role in these societies with such different values. Being designers gives us the tools to improve our way of living and understand our surroundings differently. Therefore, it is essential to have an in-depth conversation with our clients before teaming up with them. At makebardo, we always have a face-to-face meeting before getting involved in a project. It is mandatory for us to deeply understand their social responsibility and their respect and value to design. We did not start like this because we were educated to work for big companies and say 'yes' to everything for fear of not having a job. Thankfully we evolved, and we seek to keep growing without following the herd. Keeping our studio small and, consequently, our expenses low gives us one of the most important things for us—to have the freedom to say 'no'. Doing this allowed us to work closer to our clients and improve our work quality. This way of working also enables us to make wise decisions and work with people who love what they do and care to be sustainable. We appeal to the economy of resources because the fewer elements we put in our designs, the more we know what we want to say.

To finalise, we do not have the solution for the debate that we open, and in fact, that open discussion is the goal of this article. The only thing that we trust that the answers we will found in a transdisciplinary way. In the meantime, asking ourselves what we are doing to improve the world is a must and is the kickoff to solve it. We are at a breaking point in history where packaging is a big issue for our planet, so we have to take it seriously. As designers, we could go further than a double-use pack or recycled materials. There is no more time, so we must collaborate and work together to design a better world for future generations.

Bren Imboden and Luis Viale
at makebardo

'Design in every aspect needs to think ahead and lead the change; that is the most crucial task today.'

PACKAGING WITH CREATIVE STRUCTURES

As one of the packaging essentials, packaging structures play a significant role in the process of packaging. A good packaging structure infuses new blood into the product and sparks consumers' interest to explore the product it conceals. This section discusses the basic functions of packaging structure and showcases innovative and delicate packaging design. Die-cut patterns are included for specific projects.

Three Key Functions of Creative Structural Packaging Design

Packages are not simply containers of products but could also bring consumers convenience and surprises if designed with thoughtfulness on their structures. Based on three functional aspects—protective, portable and interactive—of the structures in packaging design, the following paragraphs elaborate on how designers play with the structures and make the packages helpful and fun, providing guidelines to who look for improvements in structural packaging.

'Blend' designed by Sarah Johnston.

1. Protective Function of the Structures in Packaging

One of the most fundamental functions of structures in packaging is to protect the contents of the products. Usually, structures are designed to fit the shape of the contents and, at the same time, have them fixed on the packages to prevent them from moving and bumping into each other or other objects during transportation, which is especially significant for fragile articles. Also, the spaces that structures with angles create for the contents offer them buffers or can be filled with styrofoam, newspapers or bubble wraps to further protect them from being broken. For instance, Sarah Johnston, in her project 'Blend', created trapezoidal bodies to carry candles contained in porcelains, which not only differentiate from common cuboids but also perfectly embrace the semi-cylindrical porcelains, leaving gaps between the boxes and contents that weaken the impacting force from the outer collision.

2. Portable Function of the Structures in Packaging

Another important function of structural packaging is to make the products easy to carry, saving the consumers from overloaded shopping bags. To achieve this, two main points need to be considered. For one thing, the structures must be firm enough to support the contents and keep them safe from falling during the trips back home after shopping. For another, the structures are supposed to offer the consumers a comfortable experience

'Aesop Ivory Cream Artisan Set' designed by Kenneth Kuh.

'A SCENT Essential Oil' designed by nomo®creative.

'GYVAS MEDUS Packaging' designed by Evelina Baniulyte and Gabija Platukyte.

of carrying. Specifically, the handles of the packages should be in shapes that are easy to hold and catch and should not be so sharp that might press against our hands. For example, Kenneth Kuh thoughtfully designed portable packages for the fragile 'Aesop Ivory Cream Artisan Set', of which the structures were coordinated with the products, making it possible for the consumers to carry the whole set of tableware at a time as well as for the contents themselves to arrive home safely.

3. Interactive Function of the Structures in Packaging

Rather than of 'practical use', the third so-called 'function' is more of a bonus point that turns the process of opening the packages into an interesting experience of interaction between the consumers and the products. To win this bonus point, designers should consider the messages that the brands and the products wish to convey and communicate them through visible and tangible structural packaging. Take nomo®creative's project 'A SCENT Essential Oil' as an example, the key message of this essential oil is its fragrance. Therefore, the structure was designed to 'bloom' like a flower when it is opened, envisioning the spreading of the aroma. Another example is 'GYVAS MEDUS Packaging' designed by Evelina Baniulyte and Gabija Platukyte. This package of honey opens like the wings of the bees, complemented by interior patterns resembling the veins on bees' wings. These unique structures not only make the unwrapping more playful but also reinforce the connection between the consumers and the brands.

In conclusion, to start designing the structures of the packages, three basic aspects could be taken into consideration—protective function, portable function and interactive function. Of course, there are more possibilities in the world of design where designers can carry out various explorations and experiments.

A SCENT Essential Oil

Design Agency: **nomo®creative**
Art Direction: **Chi Tai Lin**
Creative Direction: **Yu Chien Lin**
Graphic Design: **Yu Chien Lin, Wun Siang Huang**
Project Management: **Yu Chien Lin**

A SCENT is positioned as an innovative and bold lifestyle brand covering three major series of essential oils, organic care and fragrance. To connect the brand image to the local culture, nomo®creative turned to the structure and architectural totem of traditional houses in Taiwan for inspiration. They integrated the image of the roof into the symbol and drew the exclusive patterns from the window and floor totems for every flavour of the essential oil. To emphasise the pure natural, simple but surprising quality, they kept the outer box in black-and-white while saving the visual patterns for the inner part. When opening the box, the customers will be surprised by the inner pattern connected with the bottle of essential oil.

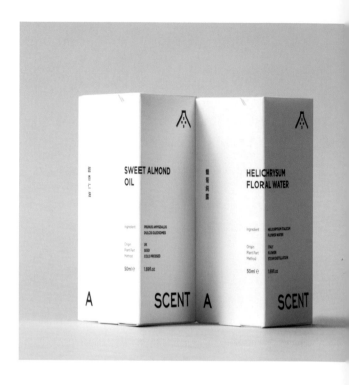

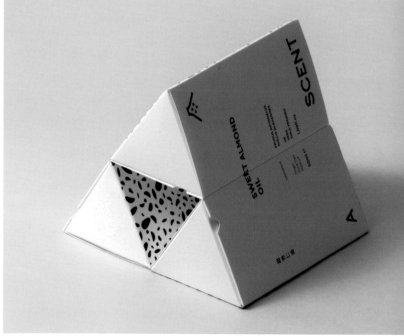

Dimensions:
Essential Oil (10 ml): 30 mm × 65 mm × 30 mm
Floral Water & Oil (50 ml): 66 mm × 66 mm × 125 mm

Material(s):
Scented paper produced by
DAYA Papers Co., Ltd. (250 gsm)

Typeface(s):
Gotham

Print & Finishing:
Planographic printing,
embossing

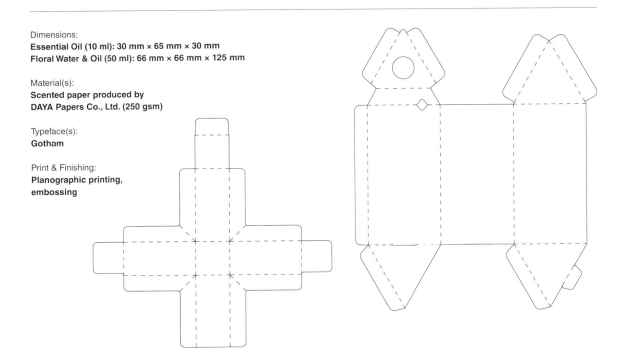

EGOME Soap Packaging

Design Agency: **K9 Design**
Creative Direction: **Kevin Wei-Cheng Lin**
Design: **Kevin Wei-Cheng Lin**
Photography: **Férguson Chang**
Client: **EGOME**

EGOME is engaged in the production of soap and in an attempt to repackage the brand image of 'pure'. K9 Design proposed a simple concept which was to launch and target the youngest market. White is the main colour of the packaging design because the design agency believed that it would be the best way to present the value of the product— purity and simplicity. For the finishing details, they added some layers to certain parts of the box through hot-stamping. Particularly, the holes on the packages were designed to present the colour of each product, allowing the consumers to have a quick glimpse.

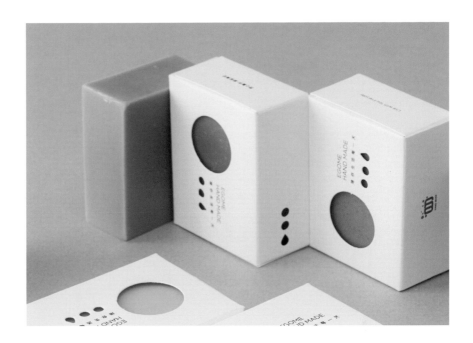

Dimensions:
82 mm × 62 mm × 32 mm

Material(s):
**Paper of Original Arjowiggins
UK series**

Print & Finishing:
Hot-stamping

GIFT PACKAGE:
HOUJYU (Kabura-Zushi)

Design Agency: **RHYTHM INC.**
Art Direction: **Junichi Hakoyama**
Design: **Junichi Hakoyama**

RHYTHM INC. created the mark that
associated with turnips which were main
ingredients of the product and devised
a way to make the mark symbolically
expressed when the upper and lower boxes
are combined.

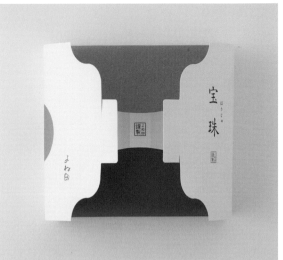

Dimensions:
116 mm × 116 mm × 34 mm

Material(s):
Paper

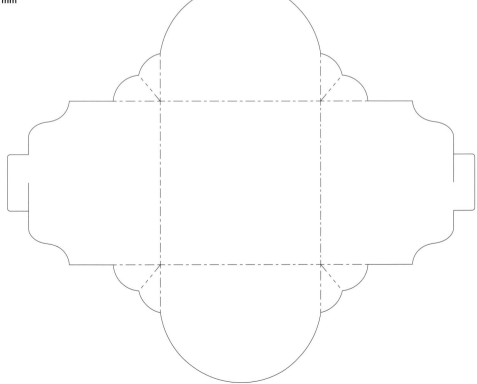

Xue Tan

Design: **Wing Yang**

Xue Tan is a health product targeting the young generation. From its name and packaging, Wing Yang aimed to make it fashionable and cool, breaking the preconception of health products consumed mainly by the elder. Meanwhile, to emphasise the nutrition of it, the designer added warm colours to complement the concise, modern and stylish visuals.

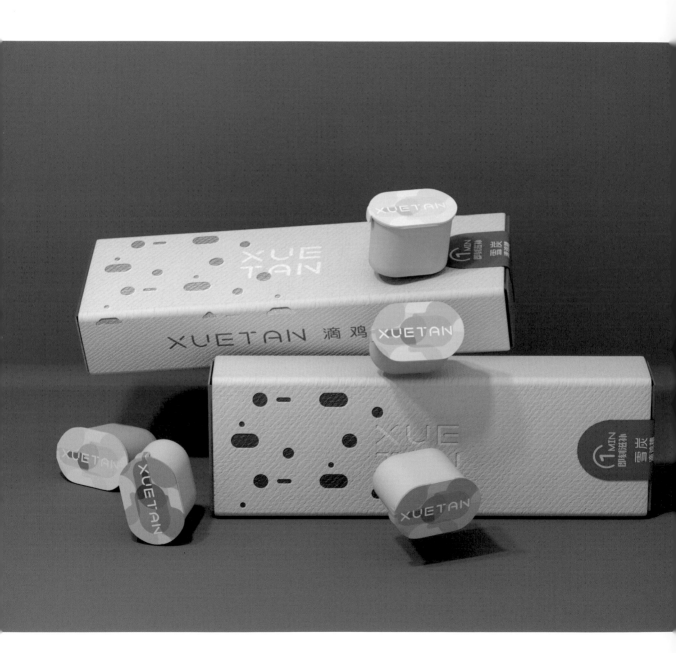

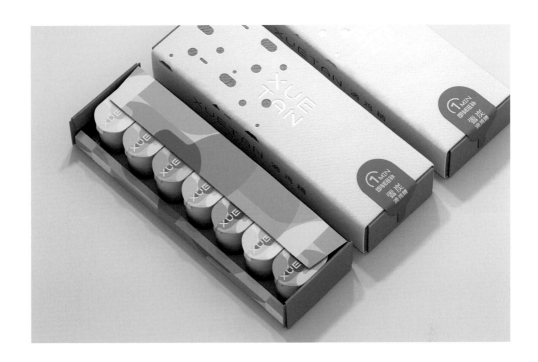

Dimensions:
100 mm × 242 mm

Material(s):
Cardboard (2–3 mm thick)

Typeface(s):
Custom type

Print &Finishing:
**UV varnish on
the logos and
geometric shapes**

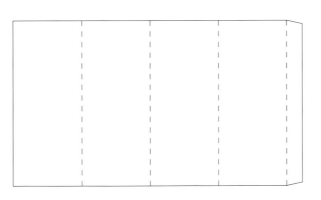

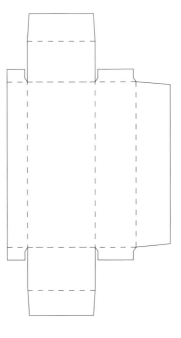

Cool Christmas

Design Agency: **Bessermachen**
Client: **Karamelleriet**

The cool and tasty Christmas calendar from Karamelleriet is filled with delicious handmade caramels. The design agency found inspiration for the design in the thought of a snowy winter and white Christmas. The products are named after a popular Danish poem about the snowman Mr Frost who falls in love with Miss Thaw.

Dimensions:
400 mm × 400 mm × 60 mm

Material(s):
Cardboard, plastic

Typeface(s):
Brandon

Print & Finishing:
**Glossy varnish and varnish
with glitter**

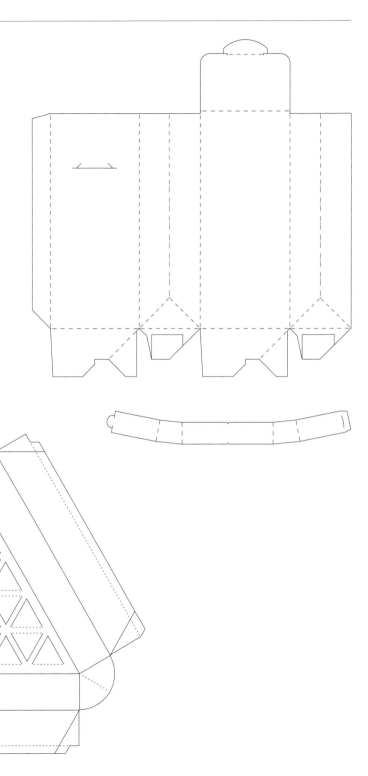

Gac Day

Design Agency: **Bratus Agency**
Creative Direction: **Jimmi Tuan**
Graphic Design: **Ton Bui, Huy Nguyen**
Showcase: **Si Tran**
Client: **Nafoods Group**

Gac Day is a daily nutrition drink that is specially developed to relieve inner health: to protect the cardiovascular system, contribute to cellular rejuvenation and strengthen the immune system. The brand name 'Gac Day' is a Vietnamese pronunciation wordplay—a friendly and memorable name for the local market. The design concept is a combination of purity and simplicity conveyed by the use of bold and solid colours. Red represents positive energy while white reflects the harmonious and serene spirit in meditation philosophy. Bottles are made from recycled plastic suitable for easy transportation.

Dimensions:
220 mm × 220 mm × 180 mm

Material(s):
Paper, plastic

Typeface(s):
Montserrat

Chinese New Year Gift Box

Design Agency: **K9 Design, AAOO STUDIO**
Design: **Kevin Wei-Cheng Lin, Louis Chiu**
Photography: **Férguson Chang**

For Chinese New Year in 2020, K9 Design created a gift box because of the custom of gift-giving during this festival. Since 2020 is the Year of the Rat, they elaborated on the meaning of rats under the zodiac context—harvest and production—in the packaging by having a main visual of the rat playing around. They adopted the hexagonal shape to express wisdom and success in the Han culture. Having these good symbols in one, the packaging conveys the meaning of Chinese New Year—sharing joy and wishing for the future.

Dimensions:
150 mm × 130 mm × 50 mm

Material(s):
Paper

Print & Finishing:
Hot-stamping

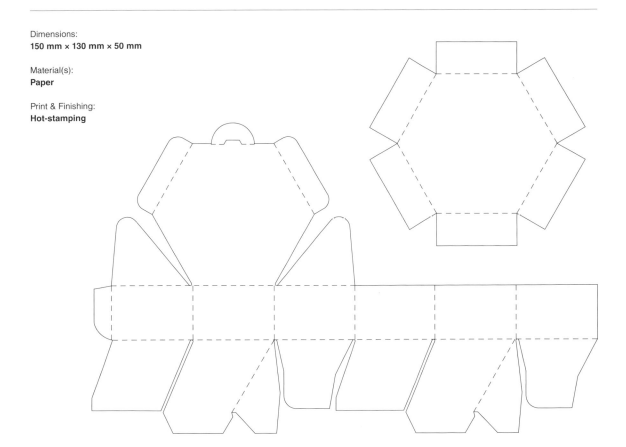

Hooked Fish Pellets

Design: **Michelle Currie**
Photography: **Michelle Currie**

The 'Hooked' packaging design is a fish pellet set that acts as a functional, sustainable alternative to traditional fish food packaging, offering a paper-based alternative to commonly sold plastic fish food containers. The colour palette and overall visual layout of the package promote a connection between the product and the user by mimicking popular pet store fish stocks and stand out among other fish food packages. Additionally, the back of each packaging piece features a daily feeding checklist. A custom typeface that imitates playful handwriting was created for the project, and the 'mouth' of each of the fish is fully functional.

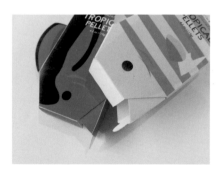

Dimensions:
130 mm × 20 mm × 50 mm

Material(s):
Cardstock

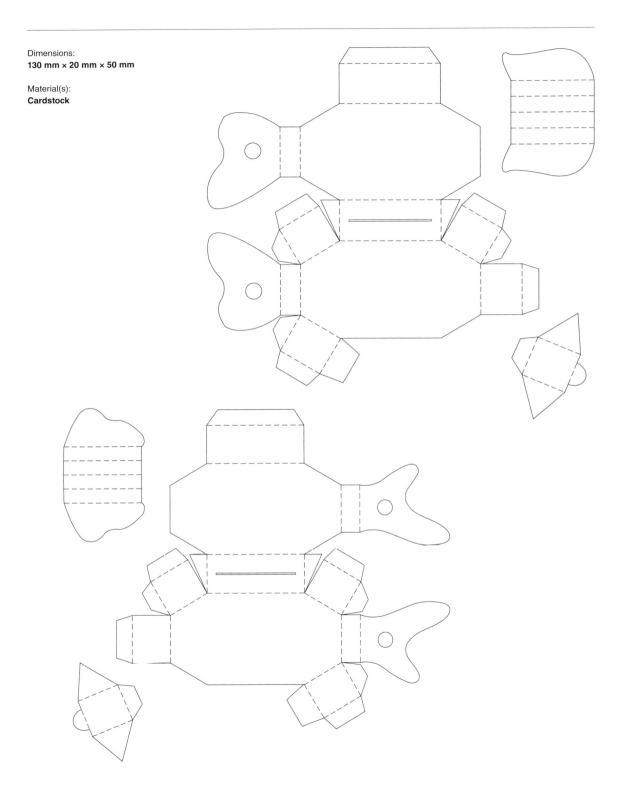

Pipas Volare

Design: **Marco Arroyo-Vázquez**
Supervision: **Ramon Abril**

Sunflower seeds is an addictive Spanish snack.
Marco Arroyo-Vázquez noticed that the main
problem of this snack is that people are too used
to throwing shells on the ground, so he decided to
develop an eco-friendly packaging for it. The grey
vines and the shape of a sunflower seed reminded
Marco of a feather. Having this association, he
settled the concept on 'birds' and based the
packaging on the idea of geometric abstraction
of a bird's dispenser. The final packaging has
two compartments—the lower has an optimised
opening for the product, and the upper works as a
container of the shells to avoid pollution.

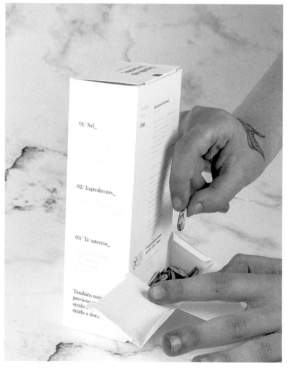

Dimensions:
170 mm × 82 mm × 50 mm

Material(s):
**Eco-friendly coated cardboard
produced by Tecknocard FSC® Mix**

Typeface(s):
Baskerville Old Face

Print & Finishing:
Offset printing

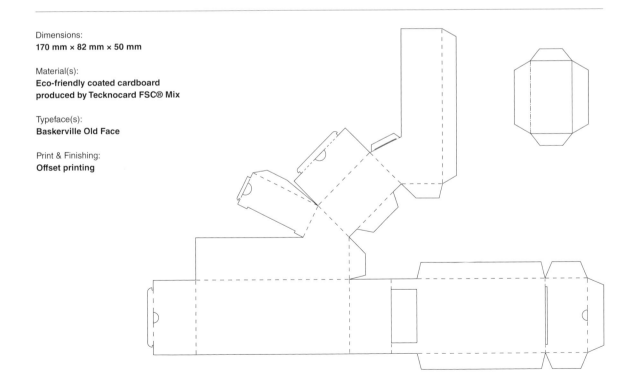

GYVAS MEDUS
Packaging

Design: **Evelina Baniulyte, Gabija Platukyte**
Photography: **Eglė Jasiukaitytė**

'Kilkų ūkis', a small family honey business, commissioned Evelina Baniulyte and Gabija Platukyte for rebranding. Firstly, they developed a unified system of boxes and paper wraps to match every size of the jars. Four types of honey are identified with distinctive colours, and the four symbols of the logotype portray the common flower of each honey. Made out of one paper folding construction with only a few gluing parts, the packaging will stand out with its unique structure. It is light yet suitable for holding the product. In the box construction, there are wings to depict bees as the symbol of honey. Besides, paper wraps were created as an economical version also with wings of bees drawn, foldable for better transportation.

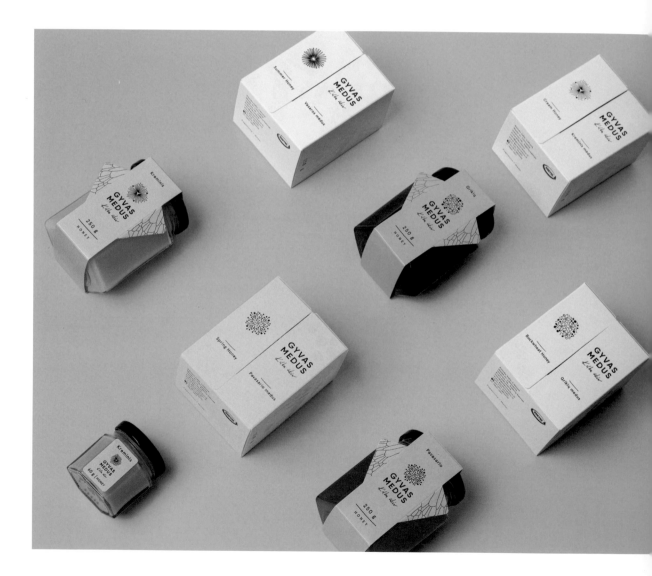

Dimensions:
500-gram jar: 90 mm × 101 mm × 80 mm
250-gram jar: 69 mm × 87 mm × 62 mm

Material(s):
Paper, cardboard

Print & Finishing:
Soft touch, UV varnish

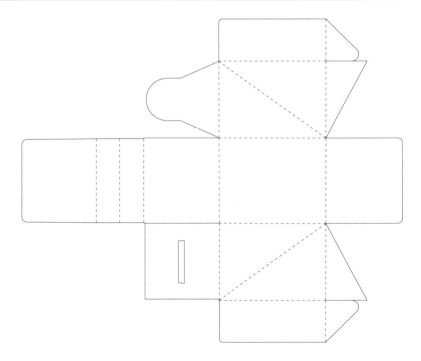

Honi

Design: **Máté Knapecz**
Supervision: **Judit Tóth, Réka Holló-Szabó**
Institution: **Corvin Art School**

The goal was a unique package coupled with a non-figurative design. The thick lines of the illustrations are honey drips that are permanent representatives of the brand. Instead of labels, there are transparent matte patterns on the jars.

Dimensions:
90 mm × 55 mm (diametre)

Material(s):
Paperboard

Typeface(s):
Comfortaa

Print & Finishing:
Offset printing

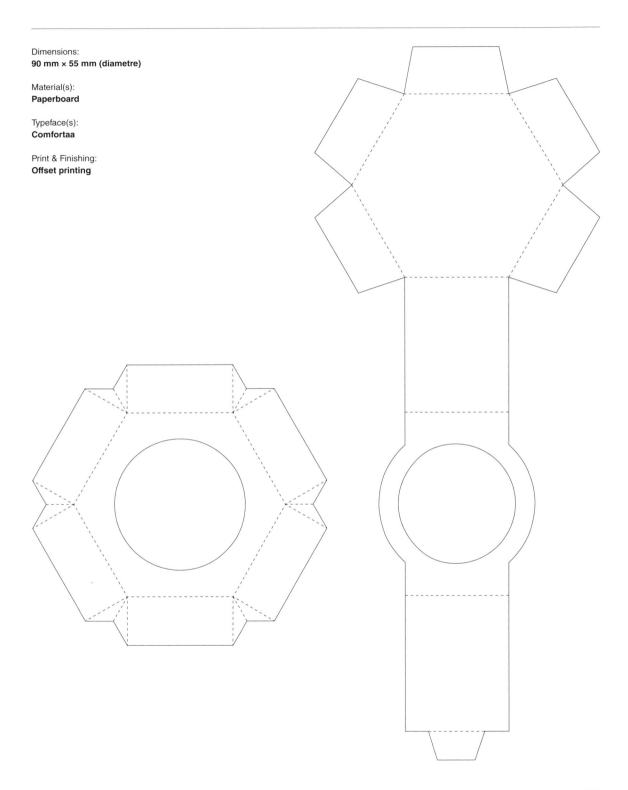

VOYAL - Elite Honey of Siberia

Design: **Ivan Gvozdev**
Client: **VOYAL**

Ivan Gvozdev was commissioned by VOYAL to design a new brand identity including the logo and packaging to express its brand values of premium quality, industry leadership and high-end quality approach. In the key image, Ivan chose a Russian bear as a symbol of power, strength and an almost fanatic love for honey. To create a widespread identity, Ivan combined white, black and gold colours and materials, giving it strong visual differentiation and recognisability compared to its competitors. For the packaging design, Ivan added a layer of cardboard to the glass jars to render it a premium touch. The cardboard boxes are in the shape of a hexagon, a perfect metaphor of honeycomb.

Dimensions:
94 mm × 70 mm (diametre)

Material(s):
Glass, cardboard, metal

Print & Finishing:
Matte lamination, hot-stamping

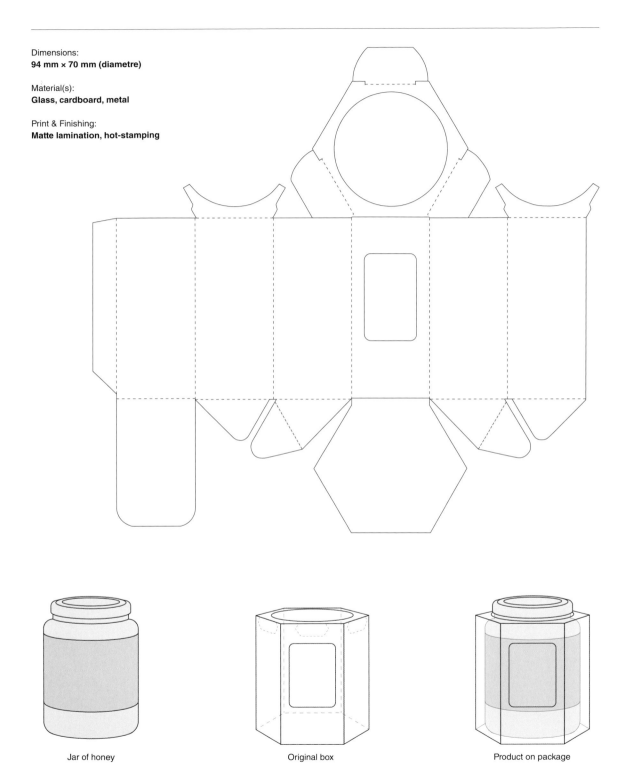

Jar of honey

Original box

Product on package

Candies Packaging

Design: **Tal Nistor**
Supervision: **Tamar Many**

This candies packaging design was inspired by Iris Van Harpen's Shift Souls collection which draws inspiration from early examples of celestial cartography and its representations of mythological and astrological chimera. The package is made in layers and simulates the movement of the garment.

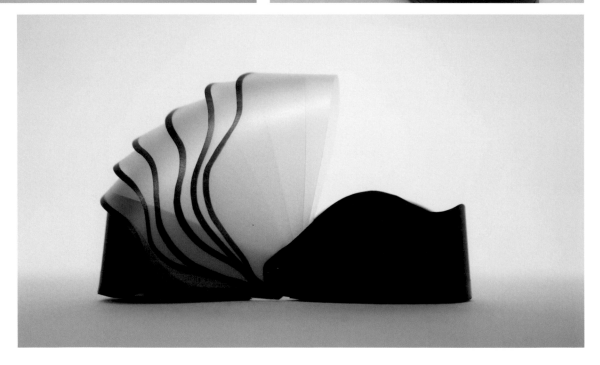

Dimensions:
115 mm × 135 mm

Material(s):
Plastic

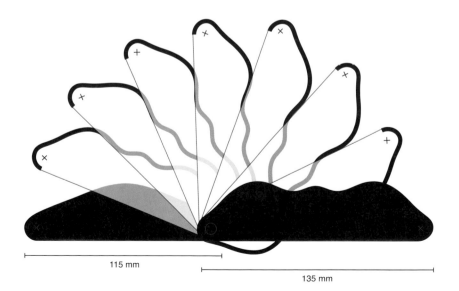

115 mm

135 mm

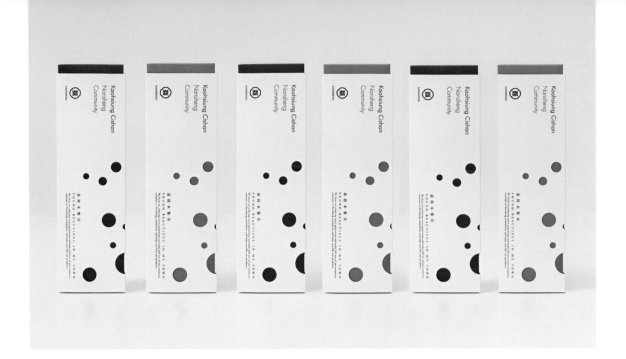

Litchi Wood Product Packaging

Design Agency: **INPIN DESIGN**
Creative Direction: **Wang Chia Ying**
Design: **Lin Chieh Hui**
Client: **Kaohsiung Cishan Nansheng Community**

Kaohsiung Cishan Nansheng Community is famous for litchis. After the harvest season, a large amount of litchi woods are left, which are recycled to make pencils and tableware as gifts for visiting and communicating events. To present the beauty of this specialty from Cishan (Ci Mountain), INPIN DESIGN made the product packages triangle like mountains. For the colours, the agency chose red and yellow inspired by litchi peels and flowers, making the products colourful on shelves to emphasise the local features and diversity. Lines and boxes are printed on the sleeve to fill information for mailing, so visitors can spread the warmth of the community.

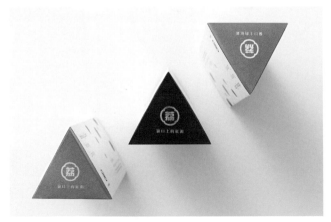

Dimensions:
240 mm × 70 mm × 60 mm

Material(s):
**Sleeve: Rasha paper
produced by
TAKEO Co., Ltd
Inner part: Lori paper
produced by
TAKEO Co., Ltd**

Print & Finishing:
Hot-stamping

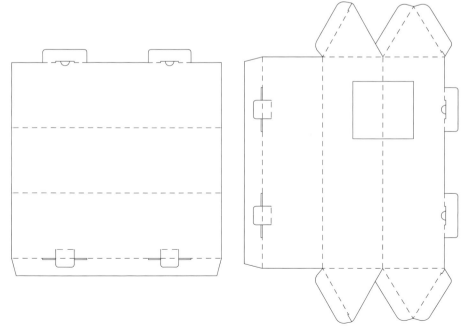

MCR Geisha Box

Design Agency: **BY-ENJOY DESIGN**
Art Direction: **Young Huale**
Design: **Young Huale, Lin Jian**
Client: **MCR COFFEE**

MCR (Micro Roast Coffee Factory) is one of the most well-known coffee roasters in China. This box of drip bag coffee, Geisha, is made of four different types of special paper. With a string tie closure, the package uses hot-stamping with silver foil, silkscreen and other printing processes, making it environmental-friendly and easy for storing. For the visuals, the bright brand colour is embraced by the understated grey, which carries a conversation between the energetic and the calm.

Dimensions:
105 mm × 105 mm × 130 mm

Material(s):
Paper, cardboard

Print & Finishing:
Hot-stamping with silver foil, silkscreen printing, white ink printing

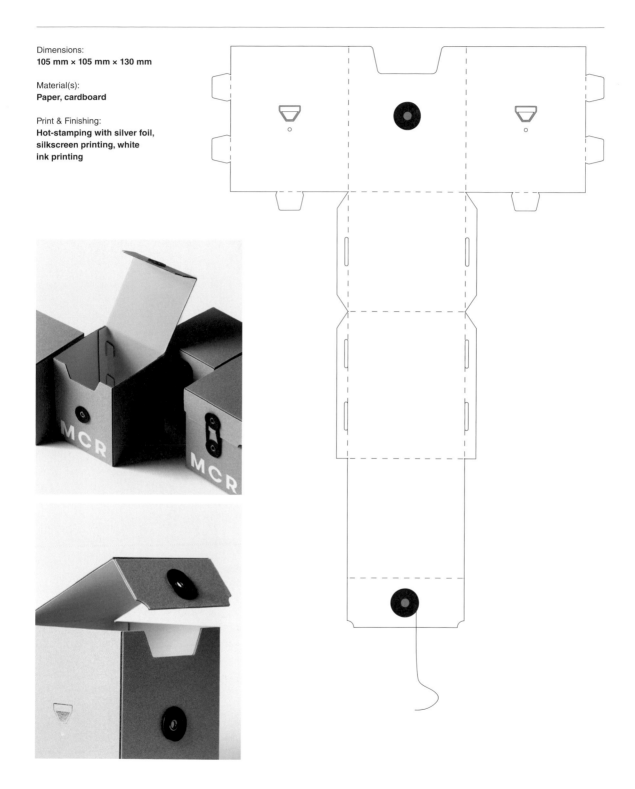

BODHA Ritual Incense Packaging

Design Agency: **STUDIO L'AMI**
Client: **BODHA**

BODHA is creating a new world of therapeutic perfumery for consumers. STUDIO L'AMI introduced new packaging for BODHA's smokeless ritual incense collection. The goal was to elevate their expanding collection with a vibrant, yet timeless and sustainable paper-based packaging that delivers a beautiful sensory experience. BODHA's mantra is brought to life through an illustrative detail as a visual reminder every time you slide open the box. Sensory details such as the blind bevel-edged emboss and tactile paper stock bring you into the moment. The box (comprised of a tray and a sleeve) offers adequate protection for the fragile cargo, through a triangular side structure that provides strength while acting as a display device for the incense.

Dimensions:
210 mm × 50 mm × 15 mm

Material(s):
FSC paper

Typeface(s):
Custom type

Print & Finishing:
Litho print onto uncoated paper, bevel embossing

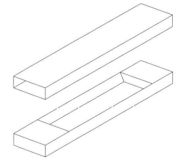

Blend

Design: **Sarah Johnston**
Photography: **Sarah Johnston,
Morgan Hancock, Paul Lee**
Client: **The Aromatherapy Company**

Blend invites customers to co-design
a candle to suit their mood or personal
preference. With six aromas to choose
from, customers can pair two to create
their own unique combination. The
narrative of divided shapes pairing
together to create a whole is echoed
in the semi-hexagonal packaging. With
a warm complementary colour palette,
the blends are represented through
tonal gradients with recipe cards on
the inside closure panels to inspire
harmonic combinations.

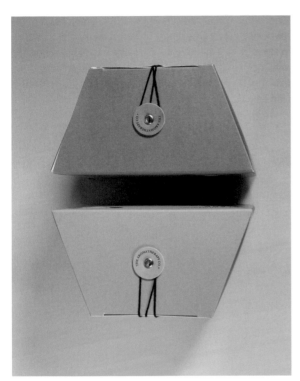

Dimensions:
140 mm × 70 mm × 105 mm

Material(s):
Uncoated card (400 gsm)

Typeface(s):
Gotham

Print & Finishing:
**Spot-colour printing
on uncoated stock**

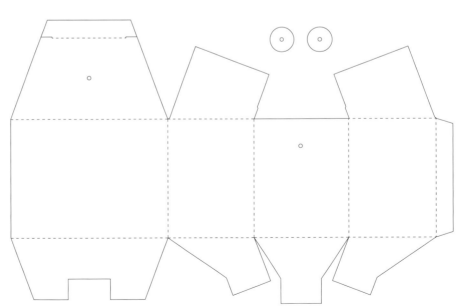

MQ COFFEE
Christmas Candy Box

Design Agency: **low key Design Company**
Design: **Chen Jieru**
Client: **MQ COFFEE**

This Christmas Candy Box contains coffee beans of cinnamon flavour produced as a special gift set for Christmas. Making it a box in the shape of a candy, the designer desired to help the consumers to share warmth and sweetness with their families and friends. Regarding the colours of the packaging, the dark green matches the Christmas vibe, and the gold foil hot-stamped on the box makes it more exquisite.

Dimensions:
55 mm × 55 mm × 130 mm

Material(s):
Art paper coated with matte lamination (350 gsm)

Print & Finishing:
Four-colour printing, hot-stamping with rose gold foil

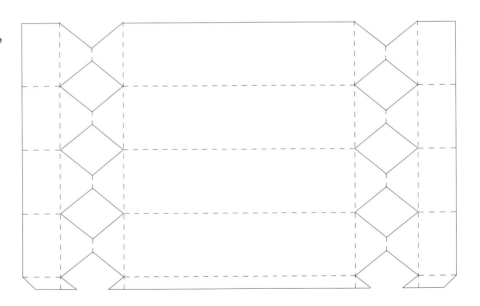

MQ COFFEE
SUPER SPECIALITY

Design Agency: low key Design Company
Design: Chen Jieru
Client: MQ COFFEE

SUPER SPECIALITY of MQ
COFFEE is a freeze-dried instant
coffee powder trying to reappear the
fragrance of specialty coffee. The
designer chose a triangular pull-out
box, making the package portable.

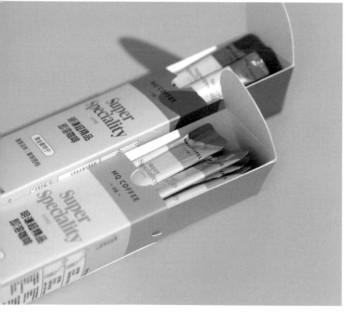

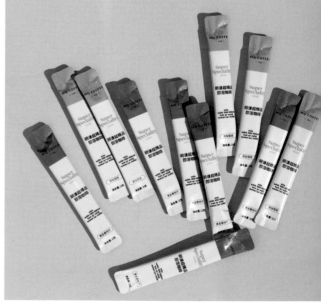

Dimensions:
55 mm × 55 mm × 130 mm

Material(s):
Art paper (350 gsm)

Print & Finishing:
**Four-colour printing,
hot-stamping with
gold foil**

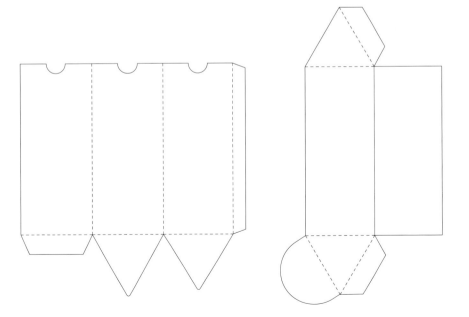

051

YUN COFFEE DRIP COFFEE

Design Agency: **low key Design Company**
Design: **Chen Jieru**
Client: **YUN COFFEE**

Named after the Chinese character ' 蘊 ' (to contain), YUN refers to the energy contained in coffee. The slogan of this set of drip coffee is 'Coffee made light', suggesting that coffee lights up our lives and makes us energetic. Therefore, the designer painted pictures of light and shadow for the illustrations of the packages, creating a peaceful and serene vibe. Also, the colours of each coffee match the flavour of it.

Dimensions:
160 mm × 55 mm × 120 mm

Material(s):
Art paper (350 gsm)

Print & Finishing:
Four-colour printing

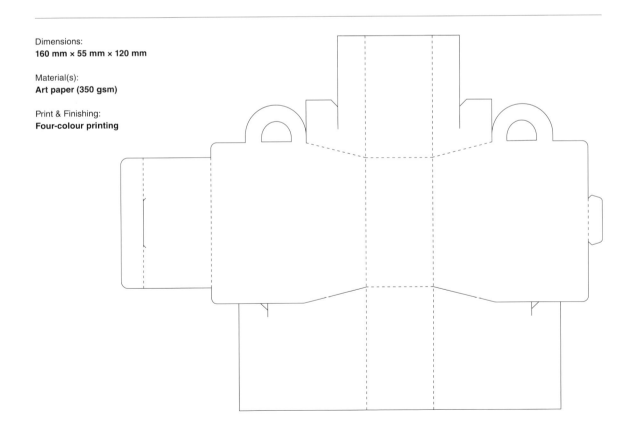

Bumi

Design Agency: **Serious Studio**
Photography: **360 Digital**
Client: **Charmaine Gonzales**

Bumi is a personal care brand that believes in making you and the world smile. Launching its self-standing bamboo toothbrushes, the brand advocates for less plastic use for a healthier self and a healthier world. While health and sustainability could be concepts that intimidate some because of many different viewpoints, Bumi allowed some sense of approachability by delivering guilt-free care starting in small ways.

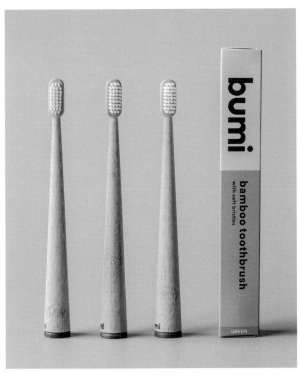

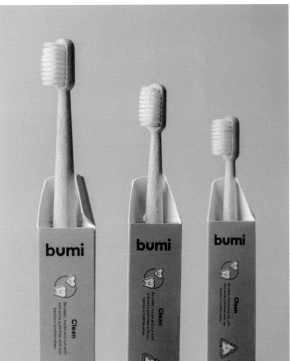

Dimensions:
210 mm × 30 mm × 30 mm

Material(s):
White box paper (250 gsm)

Typeface(s):
Circular Std Black, Circular Std Book

Print & Finishing:
Glossy finish

Mùa Cosmetics

Design Agency: **GM Creative**
Design: **Kien Kit**
Photography: **Oanh Pham, Tran Le**
Client: **Mùa Skincare**

The founder of Mùa is a soldier, scientist and agriculturist. All his life, he has protected and now is building, nurturing and protecting the natural beauty of the people and the country. Other founders of Mùa follow and act on those values.

Material(s):
Kraft paper

Print & Finishing:
Silkscreen printing

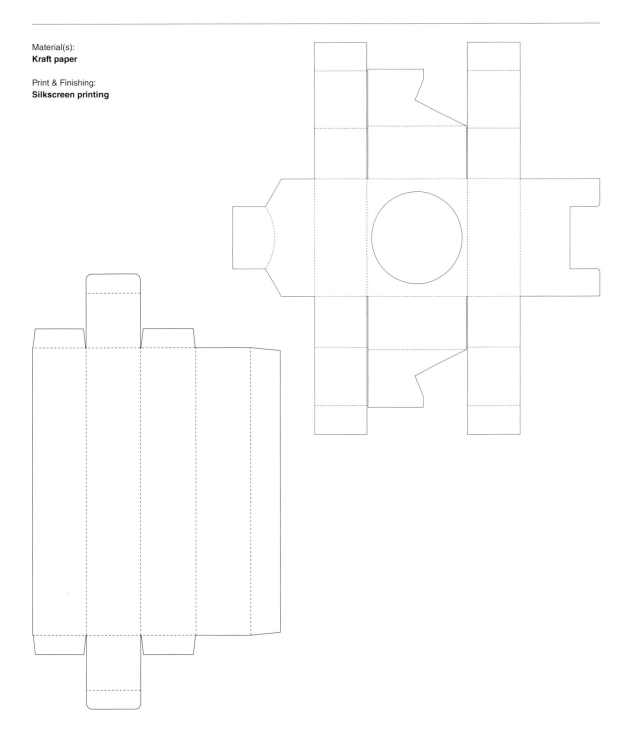

Ratri

Design: **Minsoo Kang, Jeongwon Kim**
Supervision: **Chaehyung Park**
Institution: **The Department of Design
Innovation, Sejong University**

Ratri is a premium incense brand that offers high-quality scent such as Indian made ingredients. Minsoo Kang and Jeongwon Kim defined the brand identity and designed the packages. They made the best use of the Hindu keyword to emphasise its production area of India. This concept was applied to the name of the brand and characterise each item. For instance, there are illustrations of the corresponding gods of Hinduism on the packages. The illustrations were drawn in a minimal style of lines to reinterpret and modernise gods to make the concept more accessible. Besides, the designers gave attention to the materials, using many kinds of sophisticated paper on different packages, which redound to the character of various scents.

Dimensions:
60 mm × 240 mm × 25 mm
60 mm × 80 mm × 25 mm

Material(s):
Paper

Typeface(s):
Optima

Print & Finishing:
Digital printing

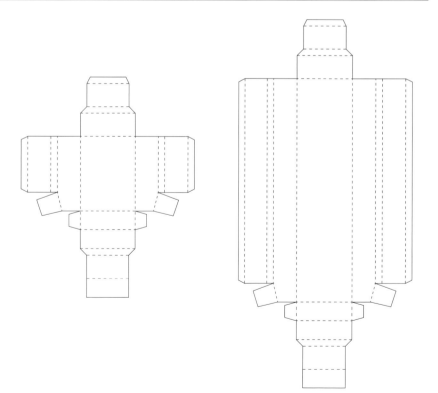

Trreeo - Soap Gift Set

Design Agency: **Victor Branding Design**
Client: **Trreeo Company Ltd.**

Trreeo is specialised in products made from Phyllanthus Emblica. In this soap gift set, Victor Branding Design avoided unnecessary packages and used paper with FSC certification to convey the brand's value of sustainable development. With an *O* representing Trreeo and illustrations of Phyllanthus Emblica, the design agency strengthened the brand identity and its characteristics on the package.

Dimensions:
205 mm × 120 mm × 35 mm

Material(s):
FSC paper

Print & Finishing:
Embossing

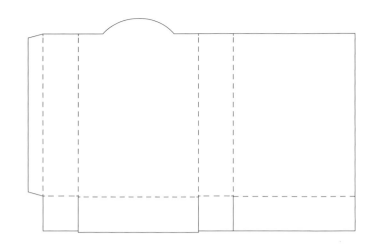

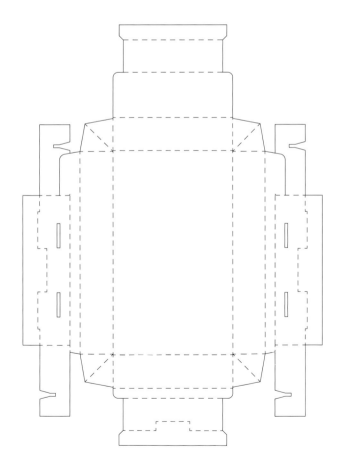

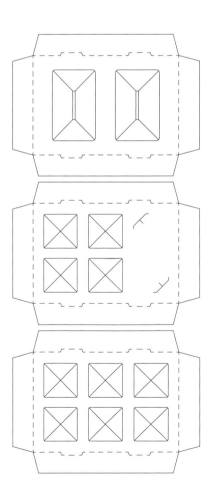

STIIK

Design Agency: **P.K.G. Tokyo Inc.**
Package Design: **Kazutoshi Amano,**
P.K.G. Tokyo Inc.
Product Design: **Eiji Sumi**
Concept: **Kenji Wada**
Client: **Ko Design Concept**

The concept of STIIK is to make a unique type of chopsticks that are like cutlery, which is suitable to Japanese food culture because it involves a fusion of different cuisines from all over the world. Usually, chopsticks are a set of two. Yet, from the exterior package of STIIK, only one chopstick can be seen. In this way, the design agency weakened the concept of conventional chopsticks and rendered STIIK similar to cutlery that works in one. Also, they designed a precise structure with a delicate assembly that makes a box containing four chopsticks possible.

Dimensions:
262 mm × 26 mm × 26 mm

Material(s):
Kraft paper

Typeface(s):
Custom type

Print & Finishing:
Offset printing

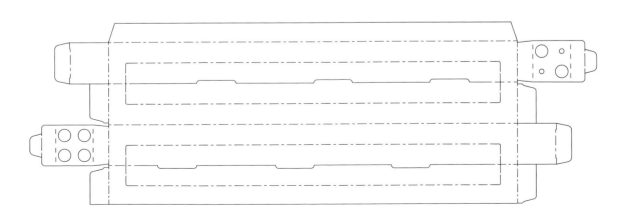

Beesweet

Design: **Bernardo Braga**
Direction: **Margarida Azevedo**

The Beesweet project promoted the creation of a packaging solution for a premium product— a drop-shaped honey container. The package is made from ecological materials, and the graphic approach is pure and minimal. White printing on recyclable kraft paper was chosen to make the colour of the product and the corresponding labels stand out. The package encloses and protects the product, highlighting honey and the belief in the preservation of bees. It has special handling with a 90-degree rotation mechanism giving access to the delicious honey. The consumers simply need to remove the strap and twist the upper top to get the product. The package is easy to assemble and, because of the folding system, it is possible to deliver the piece totally flat.

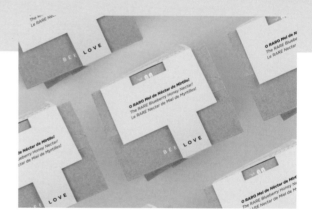

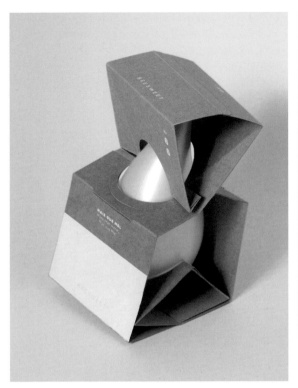

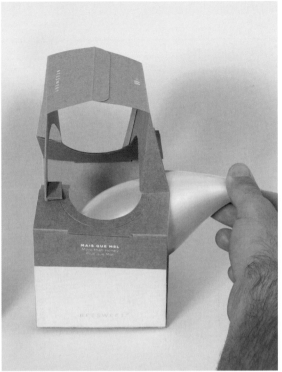

Dimensions:
90 mm × 90 mm × 150 mm

Material(s):
**Kraft paper (300 gsm),
cardboard (250 gsm)**

Typeface(s):
Gotham

Print & Finishing:
Serigraphy, one-piece cutting

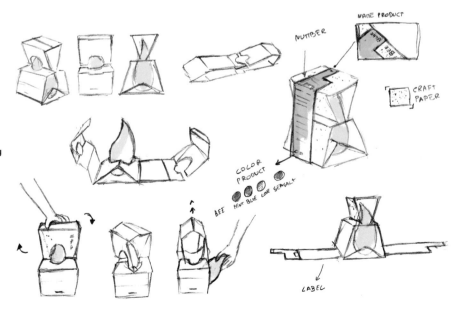

Mojelim Doctor - Hair Care Product Packaging

Design Agency: **AURG Design**
Client: **Meta Labs**

Mojerim Doctor is a hair product brand with 20 years of know-how from Mojerim, a hair care clinic specialised in hair loss. The most important thing of the packaging was to melt a high-quality image that matched the image of the top hospital within the hair caring industry into the product. Rather than simple luxury, the goal was to give the users completeness in terms of visuals and experience through delicate expression and a high level of details.

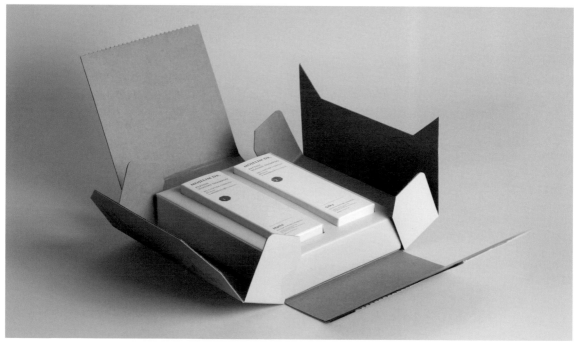

Dimensions:
52 mm × 52 mm × 170 mm

Material(s):
Paper, plastic container

Typeface(s):
Times

Print & Finishing:
Embossing

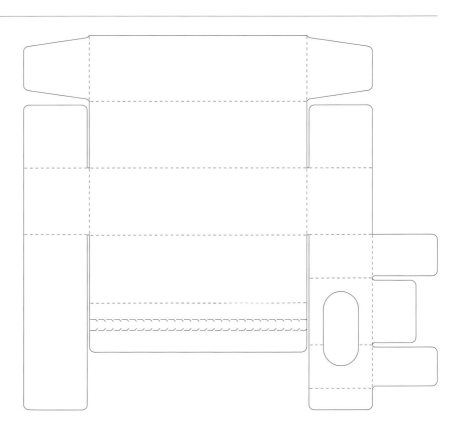

Ryabina - Home Textile

Design Agency: **Choice Studio**
Art Direction: **Aleksey Zadorozhny**
Creative Direction: **Erik Musin**
Identity Design: **Elena Astakhova**

Ryabina is a Russian brand of home
textiles of European quality, based
on simplicity and elegance. Choice
Studio's approach to the project was
the embodiment of home comfort and
family harmony. 'Ryabina' means 'rowan'
in Russian, a tree symbolising peace,
good life and a strong family bond.
Through the typographic design, the
design team tried to reflect its protective,
preserving and soothing symbolism.

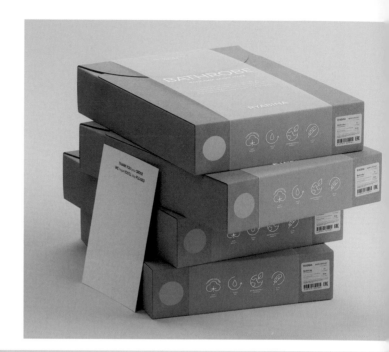

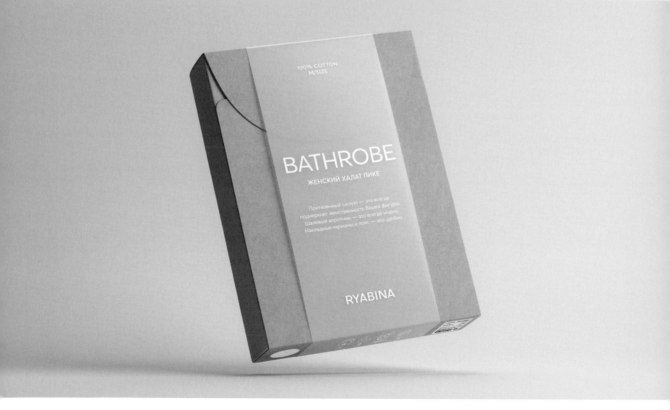

Dimensions:
420 mm × 320 mm × 70 mm

Material(s):
Carton (300 gsm)

Print & Finishing:
Single-colour printing

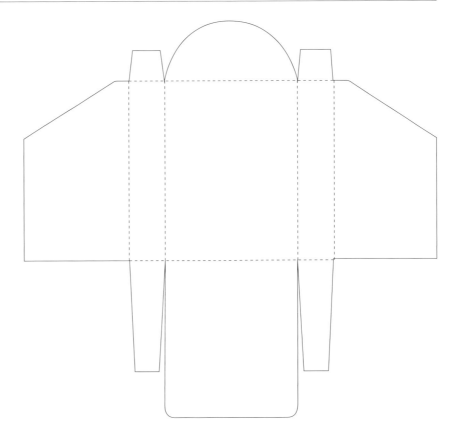

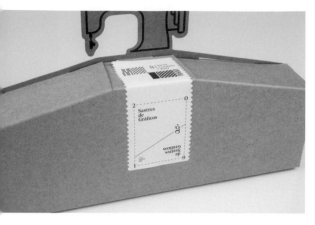
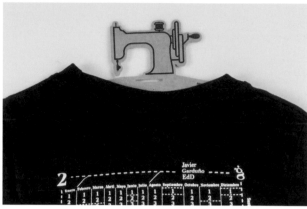

Sastres de Gráficos, Desastres Gráficos

Design Agency: **Javier Garduño Estudio de Diseño**

'Sastres de Gráficos, Desastres Gráficos' (graphic tailors, graphic disasters) is a Spanish word game that defines the design agency's activity in the studio. Based on that, they made the 2019 calendar printed on a T-shirt to simulate tailors' table of measurement. The box not only is a container of the T-shirt but can also be unfolded to be a hanger for it.

Dimensions:
220 mm × 120 mm × 90 mm

Material(s):
Carton

Typeface(s):
Didot

Print & Finishing:
Digital printing, die-cutting

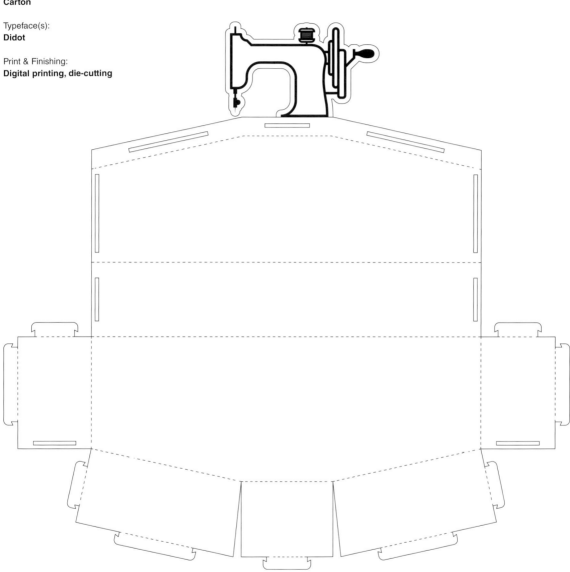

Aesop Ivory Cream Artisan Set

Design: **Kenneth Kuh**

Aesop's mission is to create a range of superlative products using both plant-based and laboratory-made ingredients of the highest quality and proven efficacy. Aesop's internal unity in concept and values has established a highly valuable material and organisational imprint through its care for the customers. This packaging set introduces them to a way of carrying the products home in a secured yet fashionable way and can be folded flat for storage anytime. After being used, the cardboard material can be recycled, ready to meet its next coming owner, establishing an environmental-friendly cycle and envisioning a minimalistic lifestyle.

Dimensions:
Mug structure:
101 mm × 108 mm × 235 mm
Plate structure:
203 mm × 89 mm × 279 mm
Set structure:
229 mm × 203 mm × 279 mm

Material(s):
Cardboard, acrylic,
Neenah Paper

Typeface(s):
Optima,
Helvetica Neue

Print & Finishing:
Inkjet matte on
Neenah Paper,
laser cutting

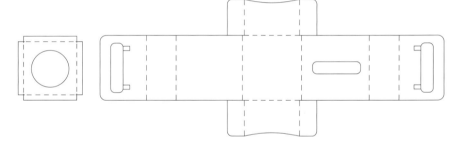

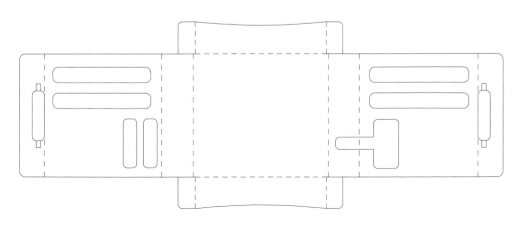

Nevera Llena

Design Agency: **ODDROD**
Design: **Andrés Guerrero**

The goal was to have a pack that would allow 14 meals to be sent to cover a person's nutritional needs for a week. After an analysis of the product, of the market and the final consumers, an identity was devised, and aspects to be highlighted and included in the packaging were defined. The packaging was designed with materials resistant to the shipping of refrigerated goods, with such dimensions that could fit within a standard refrigerator, and with a layout that would facilitate the organisation and nutritional information of meals.

Dimensions:
450 mm × 200 mm × 180 mm

Material(s):
Cardboard, paper

Typeface(s):
Mirador

Print & Finishing:
Flexography, offset printing

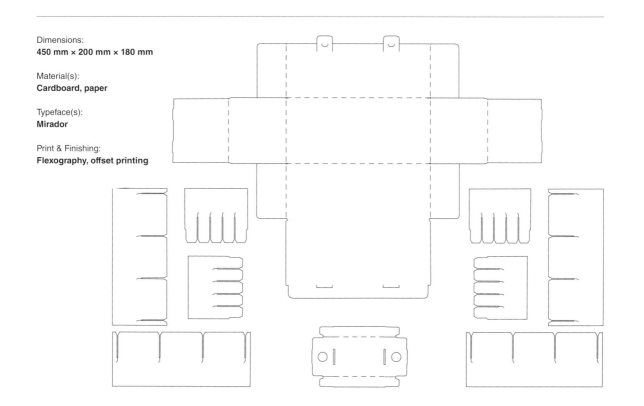

Pistacchio Cake Packaging

Design Agency: **K9 Design**
Creative Direction: **Kevin Wei-Cheng Lin**
Design: **Kevin Wei-Cheng Lin**
Photography: **Férguson Chang**
Client: **Pistacchio**

Pistacchio is an Italian ice cream shop dedicated
to presenting handmade ice cream that maintains
the original taste of the raw ingredients. Since
the products of ice cream have been developed
smoothly, the brand started to extend its business.
K9 Design created the first package of the
Baumkuchen, combining the brand's image with
concise typography and a fun logo to give the
package an elegant but playful sense.

Dimensions:
190 mm × 190 mm × 190 mm

Material(s):
Cardboard

Print & Finishing:
Single-colour printing

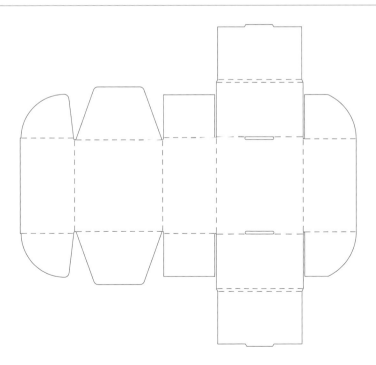

VOVE - Mosquito Incense

Design Agency: **Two Nice Studio**

VOVE is a 100% natural ingredient mosquito incense brand. With natural ingredients that would not harm health, this fragrant and pleasant scent has a diverse range of fragrances such as lemongrass, orange peel, coffee and mint. Two Nice Studio wanted to create an eye-catching, youthful and environmental-friendly package for the brand. Therefore, they used recyclable materials, especially natural ingredients, and drew playful illustrations on the packages. Besides, the outer carrier not only makes the product easily portable but also allows the illustrations to be seen from the outside.

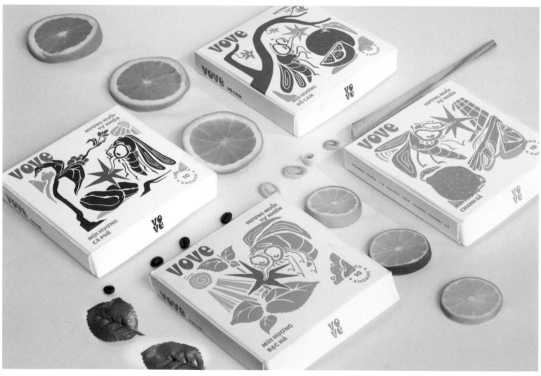

Dimensions:
Box: 130 mm × 130 mm × 30 mm
Outer carrier: 170 mm × 135 mm × 240 mm

Material(s):
Recyclable paper

Typeface(s):
Commissioner, Inconsolata

Print & Finishing:
Digital printing

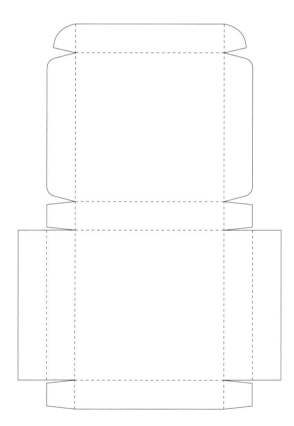

HAKKO SEIKATSU

Design Agency: **NOSIGNER**
Design: **Eisuke Tachikawa,
Ryota Mizusako, Toshiyuki Nakaie,
Ayano Kosaka**
Client: **Nishiri**

To pass on the culture of
pickles, NOSIGNER rebranded
the lineup of 'Hakko Seikatsu'
by Nishiri, a long-established
pickle company. To develop its
products and renew the long
history company, NOSIGNER
named the brand to bring
fermented foods and everyday
life together. Also, they
designed the logo using *kanji*
that can be recognised and
read easily, with bubbles as
a motif showing the process
of fermentation.

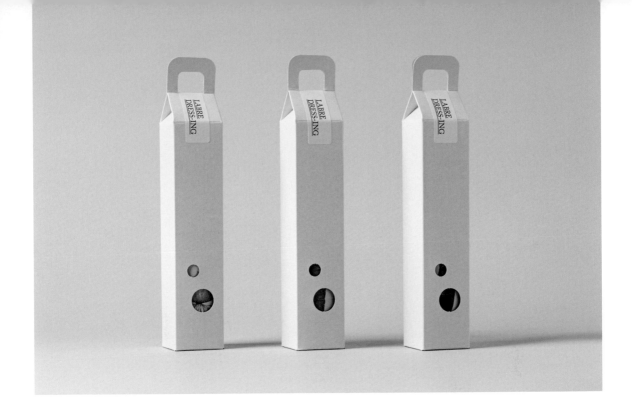

Dimensions:
Box for one: 245 mm × 40 mm × 40 mm
Box for three in a set: 232 mm × 150 mm × 42 mm
Bottle: 186 mm × 39 mm (diametre)

Material(s):
Glass, paper

Typeface(s):
Adobe Caslon Pro,
A-OTF A1 Mincho Std Bold

Print & Finishing:
Offset printing

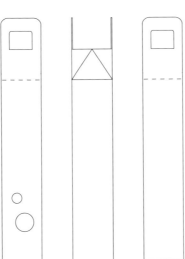

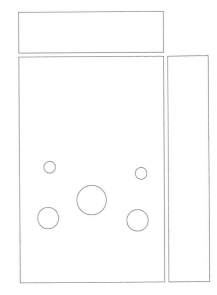

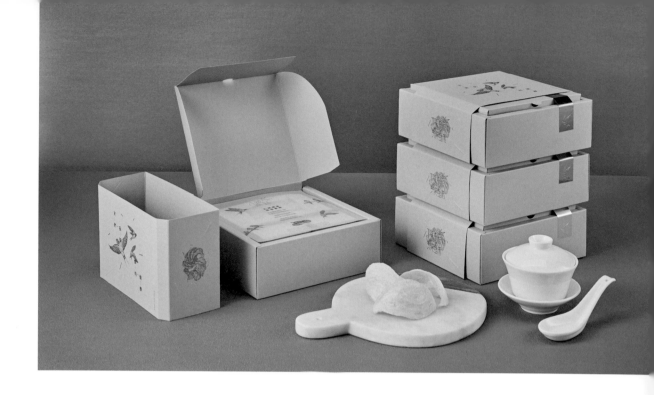

8Nests Bird Nest

Design Agency: **TSUBAKI KL**
Art Direction: **Jay Kim**
Design: **Kevin Lin, Jordan Chen**
Illustration: **Jessica Lai**
Photography: **Ferguson Chang**
Client: **8Nests Bird Nest**

TSUBAKI KL proposed an unconventional concept that would launch 8Nests and capture the modern and young Chinese market through design. Beige and grey are the primary colours as they exude an aura that people associate with being young, cool, artsy and exclusive. To harmonise the colours and the graphic assets as a shared entity, the design team looked for different ways to create the packaging box. The primary distinction that they wanted to make is between a premium item and a collectible, so they used smooth and uncoated papers and boards. For the finishing details, they decided on embossing to add depth onto certain areas of the box and hot-stamping for a classic look.

Dimensions:
180 mm × 180 mm × 80 mm

Material(s):
Antalis Popset grey card (250 gsm)

Typeface(s):
Gotham, custom type

Print & Finishing:
**Hot-stamping with rose
gold foil, embossing**

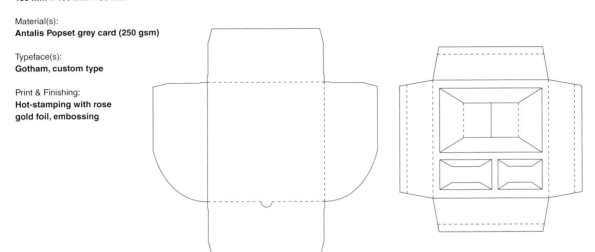

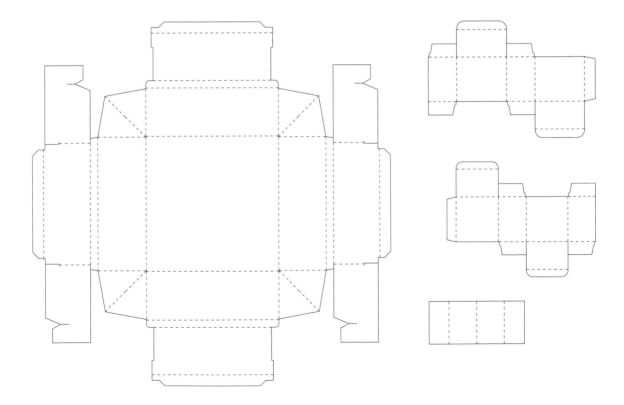

Zongzi Packaging 2019 - Dragon Boat Festival

Design Agency: **XY Creative**
Creative Direction: **Lan Hai**

Coloured in green, the outer box is fresh and concise but expressive of the festivity of Dragon Boat Festival. Special processes such as hot-stamping in gold and embossing give the package another layer of beauty, inviting the customers not only to appreciate its design but also to feel its texture. Once the box is opened, the dragon emerges together with the delicate illustrations. The design agency put a lot of efforts in details such the angle of the dragon's head, the way of folding the head inside the box and the space between each small box of *zongzi*.

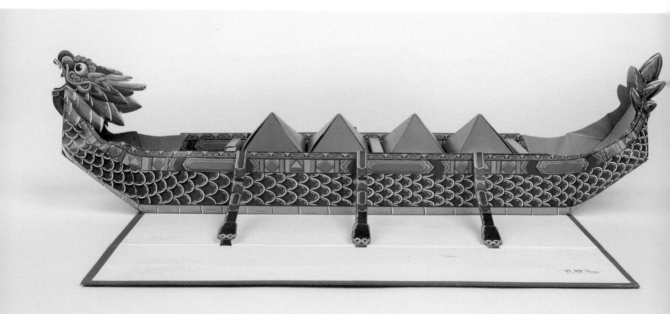

Dimensions:
470 mm × 110 mm × 100 mm

Print & Finishing:
Hot-stamping with gold foil, embossing, debossing

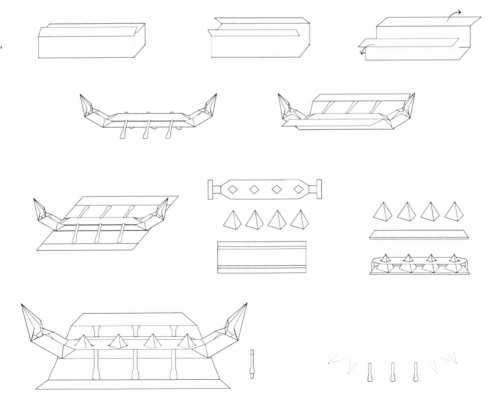

Mooncake Packaging Design 2018 - Letterpress & Poem

Design Agency: **XY Creative**
Creative Direction: **Lan Hai**

This product is inspired by the moon and Chinese poems involving the moon. The design agency combined the Mid-Autumn Festival traditions and letterpress with modern pressing processes and finishing, embedding a diversity of culture into these mooncakes.

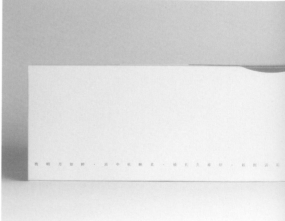

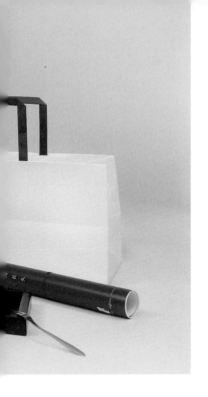

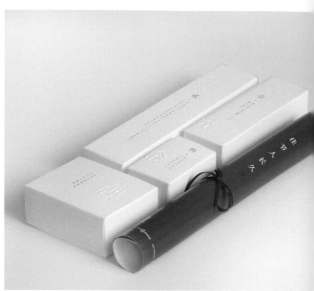

Dimensions:
270 mm × 230 mm × 100 mm

Print & Finishing:
Hot-stamping with gold foil

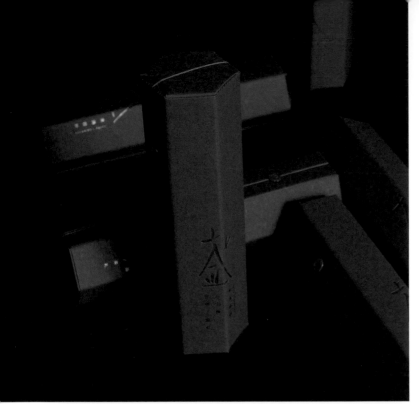

Luckiest Draw Box

Design Agency: **MEETON**
Design: **Julia Cao**
Photography: **Shawn Huang**

Drawing a fortune stick is a Chinese traditional way of the divine in temples. MEETON combined this tradition with tea packaging and created this Luckiest Draw Box of tea. The outer box is in the shape of a hexagonal cylinder which contains six triangular prisms of different types of tea representing six wishes because six is a lucky number in Chinese culture.

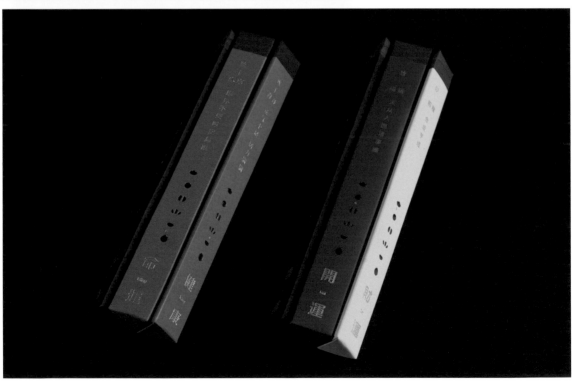

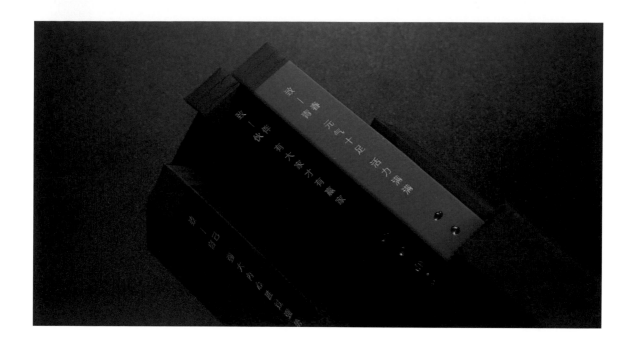

Dimensions:
55 mm × 50 mm × 190 mm

Material(s):
Boston black paper (550 gsm)

Print & Finishing:
Hot-stamping with gold foil, embossing

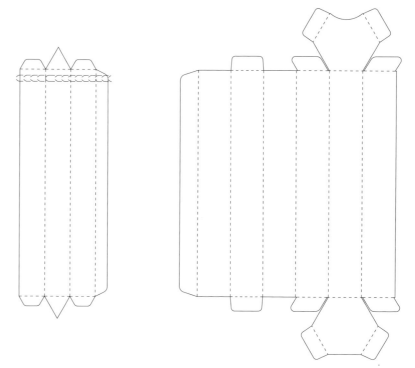

Adidas Ultra Boost X

Design Agency: **After**
General Direction: **Alfonso Fernández**
Art Direction: **Male Martínez, Andrés Villanueva**
Creative Direction: **Cástor Vera**
Photography: **Wenz Hoppus, Carlos Rojas, Alex Freundt**
Client: **Adidas**

Ultra Boost X are sneakers specially designed for women and their feet. That is why the designers decided to use the letter *X* and turned it into the main element to design the packages holding the sneakers. They also drew upon the shape to design infographics that would explain the benefits of this model. In addition, there are magnets on the sides so the consumers could easily open and close the package without damaging the product inside.

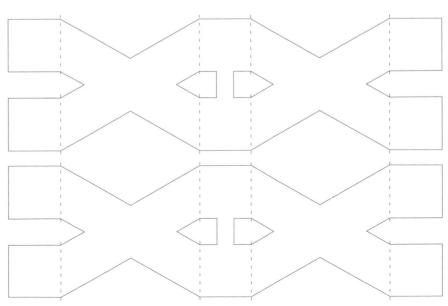

Dimensions:
Folded: 300 mm × 320 mm × 120 mm
Flattened: 300 mm × 1000 mm

Material(s):
**Paperboard Kappa - Couche
(150 gsm)**

Typeface(s):
AdineuePROTT

Print & Finishing:
Offset printing

Premium Ayu Gift

Design Agency: **Masahiro Minami Design**
Art Direction: **Masahiro Minami**
Design: **Masahiro Minami**

This is a gift package that tells the story of
ayu (sweetfish) in Lake Biwa, the biggest
lake in Japan. When the outer Paulownia
wooden box is opened, five wave-shaped
blue boxes are presented, reminding
consumers of the lake. Specially, the blue
boxes are put in the opposite directions
in turn so that they can perfectly fit into
the box without wasting any space and
can imitate the movement of waves at the
same time.

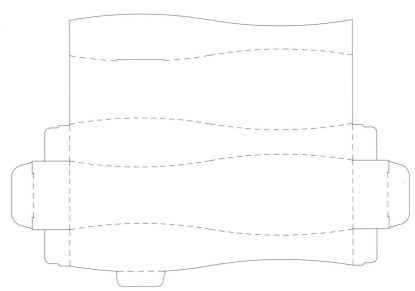

Dimensions:
Wood box: 252 mm × 310 mm × 55 mm
Paper box: 233 mm × 66 mm × 40 mm

Material(s):
Paulownia wood, paperboard

Print & Finishing:
Gold ink and glossy varnish on dark blue paper board, black and red ink on golden label

ROCCA Mid-Autumn Gift Box

Design Agency: **WWAVE DESIGN**
Design: **Amy Un Cho Ian, Ken Ho Ion Fat, Kenneth Ho**
Photography: **Andrew Kan**

The ROCCA Mid-Autumn Gift box is a set of two cookie boxes, the 'Tea Collection' and 'Floral Collection', fusing traditional Chinese mooncake with a modern European flavour. WWAVE Design tried to explore the relationship between the form, taste and cultural meaning to create a design which the consumers feel easy to relate to. They used two types of traditional paper and colour palettes for the package—English Conqueror Texture Paper in dark green for the 'Tea Collection' and Oriental Coral Earth Paper for the 'Floral Collection'.

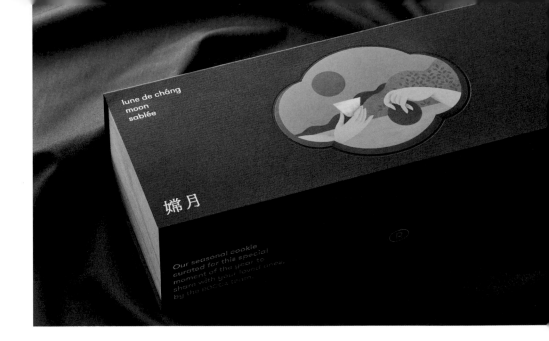

lune de cháng
moon
sablée

嫦月

Our seasonal cookie
curated for this special
moment of the year to
share with your loved ones,
by the NOCCA team.

Dimensions:
275 mm × 65 mm × 95 mm

Material(s):
**Wood, Conqueror Paper,
Earth Paper**

Print & Finishing:
Four-colour printing, embossing

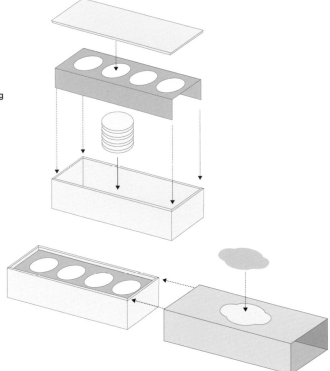

Rempah Nusantara

Design: **Vianka**
Supervision: **Deanie Cham**
Institution: **The One Academy**

Rempah Nusantara is an Indonesian company that sells a wide variety of Indonesia's native cooking ready-to-eat spices. As a brand, Rempah Nusantara embraces the authenticity of Indonesia through traditional food into instant spices. The idea is to bring the Indonesian culture through the packaging using the traditional wood carving pattern, and also the box is shaped like Javanese traditional houses due to the spices' origin of Java Island.

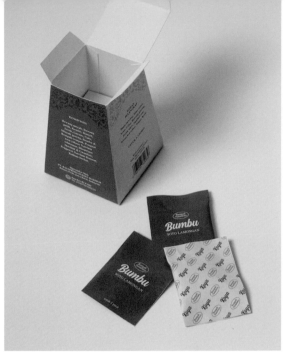

Dimensions:
Box: 65 mm × 65 mm × 90 mm
Carry box: 273 mm × 100 mm × 100 mm

Material(s):
Wudii paper, mounting board

Typeface(s):
Medallion Typeface,
Times New Roman

Print & Finishing:
Digital printing

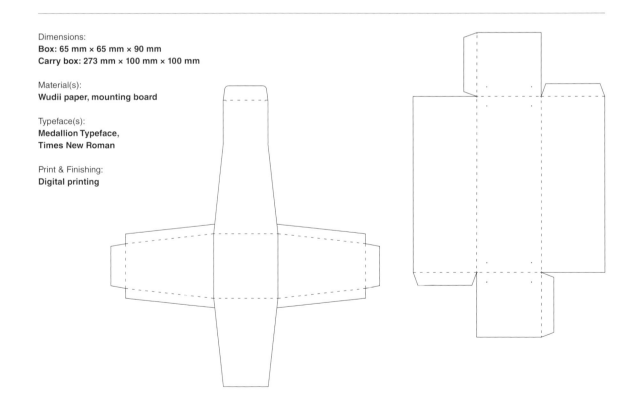

PACKAGING WITH INNOVATIVE MATERIALS

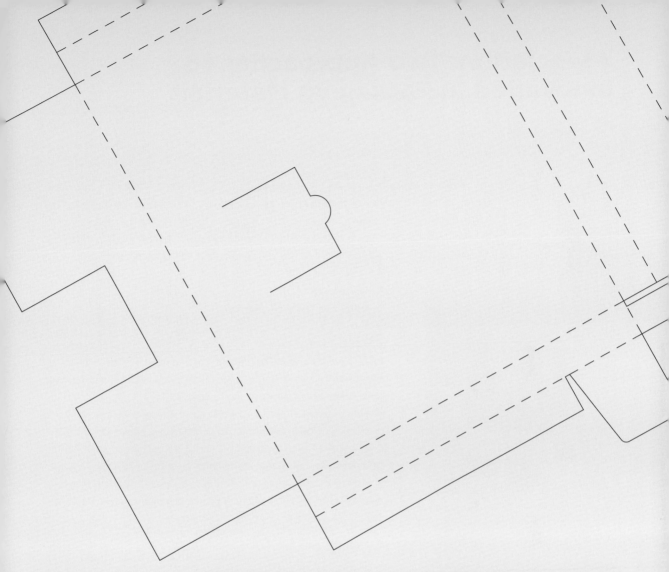

There are a lot of materials to be applied in the process of packaging, among which the most important ones are metals, glasses, papers, cartons, plastics and woods. Somehow, people tend to neglect the importance of packaging materials without noticing that a breakthrough in packaging materials will be a big surprise. This section reviews two approaches to innovation in packaging materials and provides an overview of the application of different materials in packaging.

Association: Two Approaches to Innovation in Packaging Materials

Innovation in packaging materials does not mean exploring or inventing any new substance for packaging but seeking novel ways to make good uses of the materials we are familiar with. The following paragraphs offer guidance to two approaches to innovating in materials for packaging—based on the existing connection between certain materials and products built in our heads—to play with the association and to deconstruct it.

Glass bottles designed by Zoo Studio for a wine called Costenc.

Rice bags designed by Joanna Shuen for Oryza Rice.

1. Knowing the Association

Before elaborating on the two methods, clarifying the association we have in our mind is indispensable. For one thing, we 'stereotypically' associate the contents of the products with specific materials that have been always used as their packages. For instance, eggs are usually wrapped in cartons, drinks are normally contained in glass or plastic bottles or cans, milk in paper boxes, shoes in cardboard boxes and so on. For another, we connect the contents of the products with the materials of their carriers before they are on shelves. For example, when it comes to honey, we naturally have the image of a honey comb; when it occurs to traditional Asian food, we might think of containers made of bamboos.

2. Playing with the Association

According to the association mentioned above, designers might either choose to utilise the association or to break it when deciding the materials in packaging. Playing with the association helps to reinforce the typical impression so as to convey the message of the products to the audience effectively and efficiently. As Joanna Shuen chose paper bags for Oryza Rice, the consumers would have a general idea of what the product is about without reading the texts on the packages when browsing between the shelves. Because of the texture of the paper bags complemented with hand-painting brush strokes, the packaging pleasingly communicates the naturalness of the rice which the consumers attach great importance to when

'Supha Bee Farm' designed by
Prompt Design.

'Palíndromo 5' designed by GRECO Design.

'Hard Rock Pick' designed by Marco Arroyo-
Vázquez.

buying stable foods. Another example is Prompt Design's package designed for Supha Bee Farm. The honey bottle is embedded into honeycomb core paper framed by wood which stimulates the real producing process of honey in order to present the 100% pure quality of the product.

3. Deconstructing the Association

Breaking the association leads to a feeling of alienation which brings surprise to the audience. To achieve that, there are also two basic ways that might be followed. The first is to 'borrow' a material commonly associated with a certain product to wrap another product. To be specific, GRECO Design employed egg cartons to pack a publication for the project 'Palíndromo 5' to match its theme. This fresh look would stand out in the book shops and impress the audience when they see egg cartons among the books and magazines. The second way is to replace a related material of the product with another. For example, CDs are usually carried by plastic cases, but Marco Arroyo-Vázquez applied wood to the project 'Hard Rock Pick' and enclosed the CDs of limited editions in wooden cases, overthrowing the former association yet rebuilding the connection between music and the product through the shape of a guitar pick.

To sum up, association might be a helpful starting point for designers who are selecting materials for their packages. Based on the knowledge of all those associations between the products and the materials, they are able to consider approaching the materials by reinforcing the association or by overthrowing it. In either way, their ultimate goal is to design a visually and texturally impressive package that tells the brand's story.

Well, K. Naturally Flavored Mints Packaging Design

Design Agency: **milieu studio**
Art Direction: **Xuezhou Yang, Shanshan Chen**
Design: **Xuezhou Yang, Shanshan Chen**

Concerning designers' health problems, milieu studio camp up with the idea of developing a portable, chic and minimalist packaging for digestive tablets. To distinguish the tablet packaging from a bulky cylindrical and text-heavy white plastic bottle commonly found on the market, the design agency created a slim octagon shape that can be slid into bags and pockets of any size and prevent itself from rolling around on a tabletop. Contrasting with the effective and photogenic primary package, the secondary one is rather simple—a single sheet of thick paper with one fold merely enough to hold the primary packaging and to list all the necessary information.

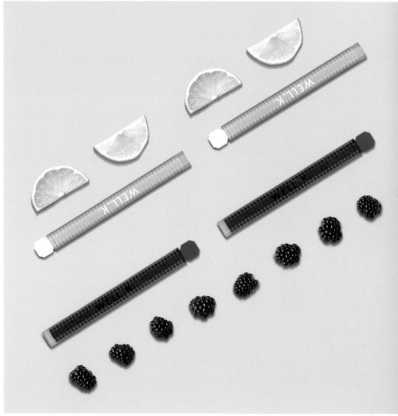

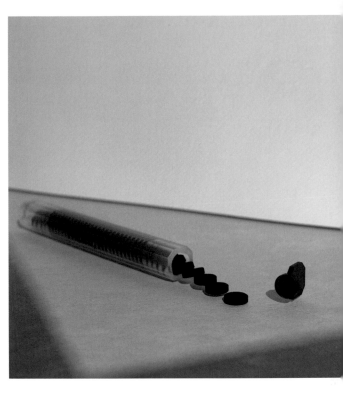

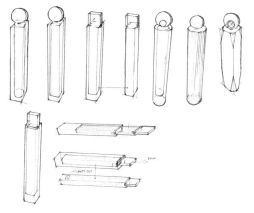

Dimensions:
Primary packaging: 14 mm × 14 mm × 130 mm
Secondary packaging: 325 mm × 3 mm × 60 mm

Material(s):
Primary packaging: translucent injection-moulded acrylic
Secondary packaging: uncoated paper (486 gsm)

Typeface(s):
PingFang font

Print & Finishing:
Digital printing, silkscreen printing, matte finishing

The slim octagon-shaped package is produced with translucent acrylic that is injection-moulded. Its special shape helps itself to stay on a tabletop steadily, and its matte surface gives itself a unique touch.

El Tresor

Design Agency: **Zoo Studio**
Art Direction: **Gerard Calm, Jordi Serra**
Design: **Jordi Serra**
Client: **Set & Ros**

El Tresor (the treasure) is a superlative organic oil made from the finest olives. The challenge of this project was to give personality and exclusivity to this little bottle of 50-millilitre oil, creating a special piece that would impress the consumers. With these premises, Zoo Studio created an ephemeral packaging to be destroyed by the consumers before they enjoy the content.

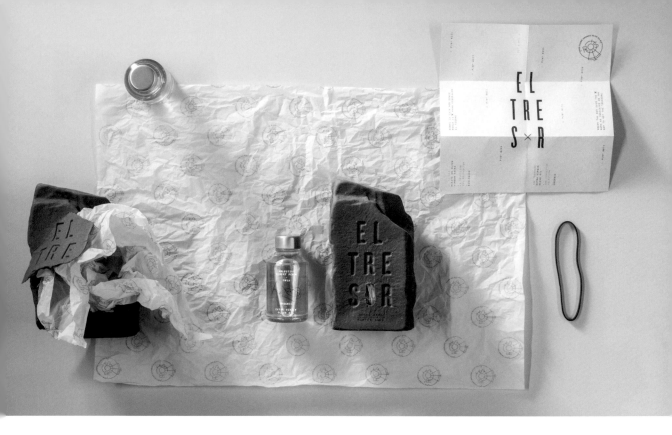

Dimensions:
Outer package: 80 mm × 70 mm × 140 mm
Bottle: 80 mm × 40 mm (diametre)

Material(s):
Paper pulp, glass, silk paper

The outer shape of the package was inspired by 'Ilicorella', a type of slate stone native to the Mediterranean lands where the olive trees grow. It was produced with recycled paper pulp, an ecological material that transmits roughness but is easy to open, mimicking the action of digging the soil.

AJOTO the Pen Packaging

Design Agency: **AJOTO**
Design: **Christopher Holden, Marta Verdes Montenegro Diaz**

Aiming for making the packaging more practical and engaging, AJOTO's approach was to combine thoughtful designs with sustainable materials and processes. By introducing additional elements into the mold, the box can be turned over once opened to form a pen rest—a functional piece that would sit beautifully on any desk or table. Moreover, the design agency indulged in the world of letterpress and blind embossing to create a packaging that brings together precision and craft in equal measures.

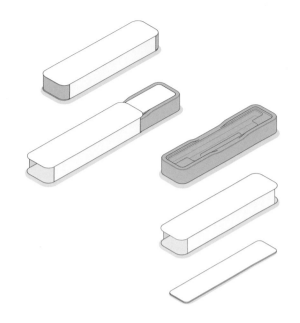

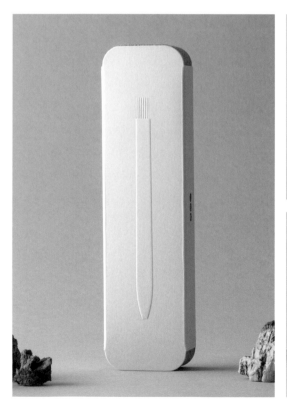

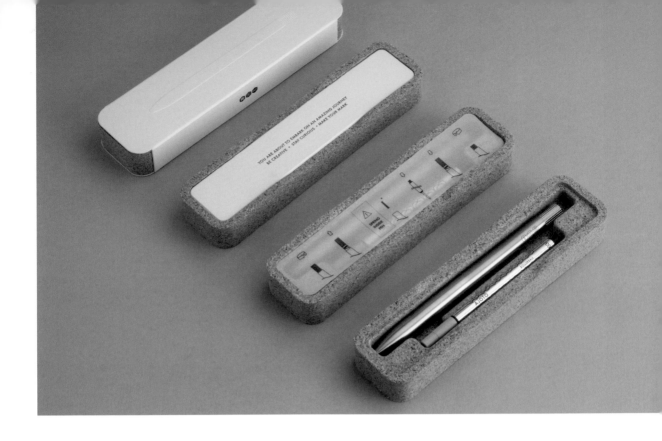

Dimensions:
160 mm × 40 mm × 20 mm

Material(s):
Cork, paperboard

Print & Finishing:
Letterpress, blind embossing

For materials, the design team chose cork, a natural, rapidly renewable material that is not only bio-degradable and recyclable but also light, anti-bacterial and very resilient. All these attributes make cork the perfect material for protecting the pen during transit.

Apotheke Fragrance
Incense Sticks

Design Agency: **UNPAC**
Design: **Taro Uchiyama**
Brand Direction: **Keita Sugasawa**
Photography: **YOKOYAMA**

Apotheke Fragrance is an artisanal, handmade fragrance brand who admires the simple elegance and analogue textures found only in the fruits of handmade labour and shuns machinery and outsourcing. The brand began solely with a single product lineup of candles and has developed a wide range of fragrance items including these incense sticks.

Dimensions:
370 mm × 26 mm (diametre)

Material(s):
Kraft paper, cork

Print & Finishing:
Embossing

To remain consistent with the brand's style of minimalism, kraft paper tubes are used to carry the incense sticks. Uniquely, these tubes are sealed by corks which are more common in metal packages or glass jars. The printing of the labels involves embossing, giving the product another layer of analogue texture.

STENDHAL

Design Agency: **Zoo Studio**
Art Direction: **Jordi Serra**
Design: **Jordi Serra**

'Atelier des auteurs' (authors' workshop) is a concept of perfumery brand that Zoo Studio designed to present the latest innovations from Pujolasos in the fields of perfumery and cosmetics. The 2018 proposal consisted of a luxury product targeting adults and presented in the Fair PCD Paris 2018. In 2019, Zoo Studio designed an elegant, sober brand, inspired by the image of Paris—a city with an important artistic and pictorial heritage—framed in the 18th century.

Dimensions:
40 mm × 50 mm × 50 mm

Material(s):
Wood

It was the perfect chance to present a collection of perfumery caps produced with painting frames. In the design of the label and the pack, Zoo Studio used textured materials to emphasise the idea of artwork.

CHOCOLATE ACADEMY
- Wines for Chocolate & Chocolates for Wine

Design Agency: **Zoo Studio**
Art Direction: **Gerard Calm, Xevi Castells**
Design: **Gerard Calm, Xevi Castells, Maria Blanch**

Zoo Studio was invited to design the packaging for several projects by Ramon Morató and Miquel Guarro, two pastry chefs from the Chocolate Academy of Barcelona, who explode their talent under the brand Cacao Barry. The chefs got inspired by varied concepts and topics to develop innovative recipes with a high gastronomic and aesthetic value. In Zoo's proposals, they wanted to design packages that would enhance and complement the attributes of the products. 'Wines for Chocolate & Chocolates for Wine' is one of the proposals.

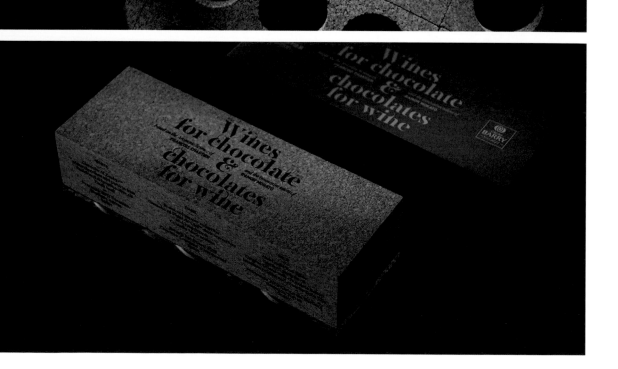

Dimensions:
70 mm × 70 mm × 200 mm

Material(s):
Cork, methacrylate

The materials used are cork and methacrylate for the base. This combination is a parallelism to a wine bottle (usually made of glass and sealed with a cork) to further illustrate the theme.

Material Matcha Uji

Design Agency: **Rice Studios**
Creative Direction: **Joshua Breidenbach**
Design: **Mark Bain, Joshua Breidenbach**
Photography: **Mark Bain**
Client: **Material Matcha Uji**

The Material Matcha Uji identity asks the viewers to contemplate pure materials. In packaging and communication, spectra of unadulterated materials are utilised to represent aspects of the product's creation. The logo mark was created to be imprinted, stamped or otherwise stenciled into almost any substrate. With this flexibility, future packaging can be made from a plethora of materials and simply branded.

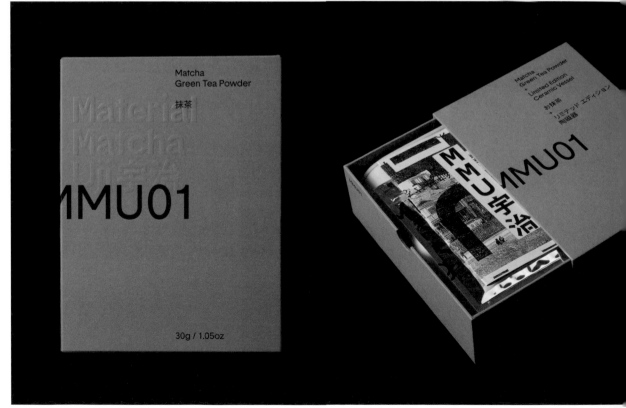

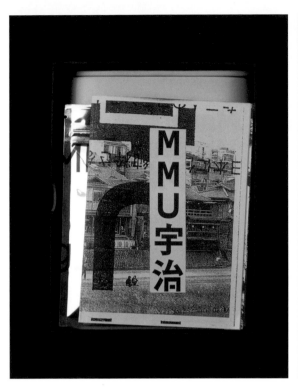

Dimensions:
165 mm × 130 mm × 40 mm

Material(s):
High-gloss Mylar, ceramic vessel

Print & Finishing:
Offset printing, embossing

Space-age high-gloss Mylar is an otherworldly material, which contains the pure matcha. A zine, made of paper, and photocopy toner are juxtaposed with the Mylar to express a handmade side to the process. A ceramic vessel, made from recycled ceramic material, represents the history and *wabi-sabi* tradition of the Japanese tea ceremony. Three outer boxes denote each product's distinct flavour profile.

Hard Rock Pick

Design: **Marco Arroyo-Vázquez**
Photography: **Marco Arroyo-Vázquez**
Client: **Hard Rock International
(Hard Rock Cafe)**

This concept project for Hard Rock is a CD Special edition package based on a 'guitar pick', an element related to the world of music and easily identifiable by anyone regardless of their nationality. It was conceived as an artistic piece, committed to the romantic of music, who appreciates CDs of limited editions. The main claim of Hard Rock Cafe is 'LOVE ALL - SERVE ALL', and the designer chose this reference and changed 'serve' to 'listen' that connected to music more tightly.

Dimensions:
204 mm × 177 mm × 30 mm

Material(s):
MD Wood, neodymium magnets

Print & Finishing:
Burning, engraving, laser cutting

Different from traditional plastic CD cases, the designer chose wood and made a pick-shaped case. In addition to being a nod to the material with which guitars are made, wood allows offering new connotations such as touch, smell and sight. To be specific, as the design is engraved and burnt on the wood, there are a unique texture and a differential smell. Also, the wooden pick-shape arouses a desire to feel and touch.

Chen Li - Playlist Album Packaging Design

Design Agency: **HOOOLY DESIGN**
Art Direction: **Cao Fan, Shen Hongrui**

Released in 2018, *Playlist* is the album of the young Chinese singer and music producer Chen Li, featuring ten songs. This entire album is designed around the concept of 'play'.

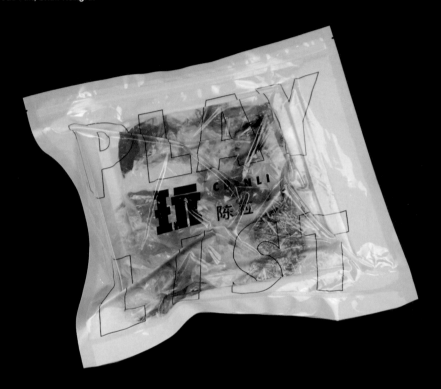

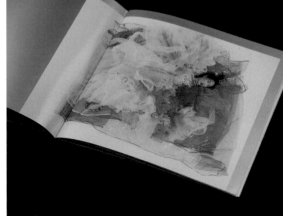

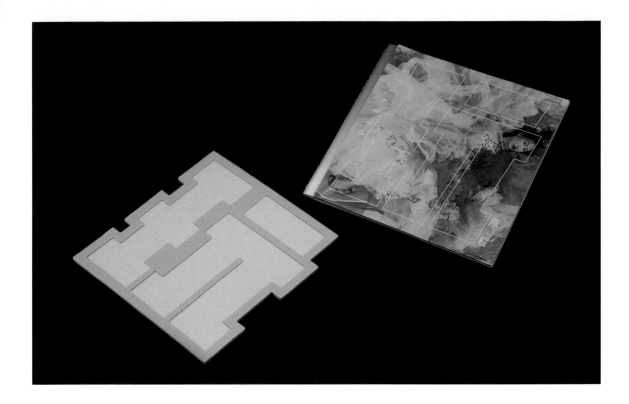

Dimensions:
330 mm × 330 mm

Material(s):
Plastic

Print & Finishing:
Spot-colour printing

Special Process:
Vacuum-sealing

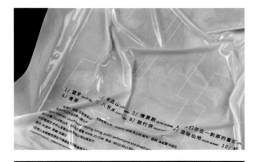

To extend the concept and coordinate with the photographer Haal's idea of plastic package, HOOOLY DESIGN employed vacuum-packing to invite the audience into an interactive experience like unpacking a toy.

3 Magazine

Design Agency: **Studio D.U.Y**
Design: **Duy Dao**

Published three times a year, 3 Magazine is a journal that explores culture and society with the idea of number 3. Each issue chooses one three-letter word, three fonts and three colours that respond to society during that period of time. The magazine looks at current social issues along with events in our everyday culture and makes a conversation around it. Using first-person interviews, as well as related articles and photography, 3 Magazine's mission is to bring another point of view to the audience.

Dimensions:
152 mm × 229 mm

Material(s):
Uncoated paper, vinyl bags, acrylics

Special process:
Vacuum-sealing

In order to package and include three physical letters for
every issue and maintain a versatile vessel for unique cases
of contents, the smartest way is to use vacuum-packaging
solutions. This resolution is not only flexible and adaptable to the
concept and the design but also cost-friendly and eco-friendly
by using bio-degradable and composable vacuum-sealed bags.
The unorthodox packaging gives the magazine a distinctive
visual representation.

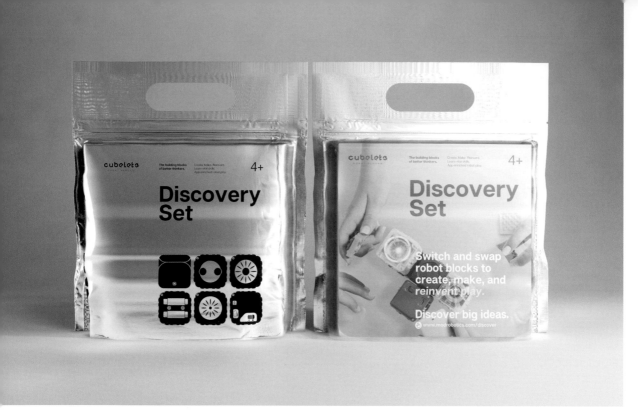

Modular Robotics

Design Agency: **makebardo**
Creative Direction: **Bren Imboden, Luis Viale**

Modular Robotics is a small but growing team headquartered in Boulder, Colorado, giving a new point of view on how to play with toys. The aim was to create a design that appeals not only from a kid-centric perspective but also attracts all ages. 'Discover' is a key factor in this toy, and the audience can connect cubes and build thousands of different robots.

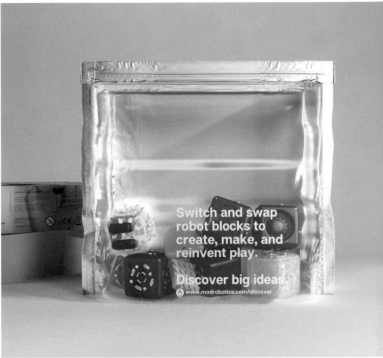

Dimensions:
214 mm × 214 mm

Material(s):
Sustainable plastic, printed cardboard

Typeface(s):
Articulat CF

Print & Finishing:
Four-colour printing, matte finishing

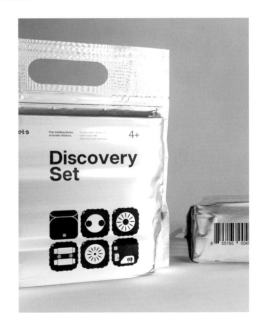

Following the concept, makebardo put the packaging inside a static shielding bag to hide the front of the box to surprise the audience and let them discover it. With this concept of a closed bag, the product also gains direct brand exposure on the streets by having a bag with handles, thus a second bag from the store is not necessary.

Maison Marou
Mooncakes

Design Agency: **Rice Studios**
Creative Direction: **Joshua Breidenbach**
Design: **Daniel Keeffe**
Project Management: **Paula Vo**
Photography: **Hieu Duong**
Client: **Maison Marou**

For Maison Marou Mooncakes this year, Rice Studios' intention was beyond creating a beautiful, practical box. They sought to design a box that someone would keep, a box that would have a second life. As mooncakes are given as gifts, considering the impression they would make in the hands of the receivers was important.

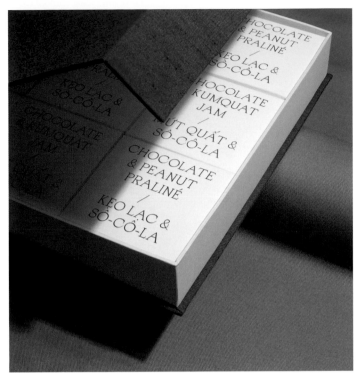

Dimensions:
White box: 90 mm × 90 mm × 35 mm
Red box: 90 mm × 95 mm × 50 mm
Yellow box: 200 mm × 205 mm × 50 mm
Blue box: 290 mm × 290 mm × 50 mm

Material(s):
Bookbinding cloth (190 gsm),
brilliant white stippled art paper (200 gsm)

Print & Finishing:
Hot-stamping with gold and bronze foils

Rice Studios used pre-made fabrics selected from a market based on the way they feel and used metallic foils only. It was part of the concept to go on a no-printing process.

PULL TEAR AND REPEAT

Design Agency: **Hansen/2**
Design: **Gesa Hansen, Johanna Flöter, Daniel Hansen**
Client: **RESET ST. PAULI Druckerei GmbH**

The calendar PULL TEAR AND REPEAT is about discovery in a playful and curious way—when the audience tears off the paper, they are creating their own calendar by repeating it each month. The remaining pieces of the cracks symbolise the past time. The calendar is in the present, what we can see is the past, and the future is locked and unknown. Therefore, the packaging provides an experience of unpacking, discovering and destroying the object.

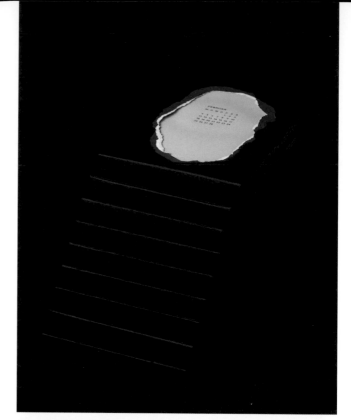

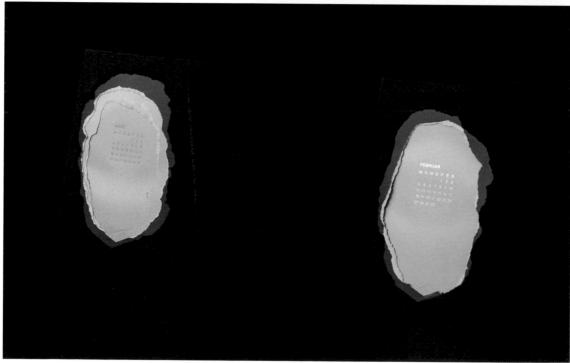

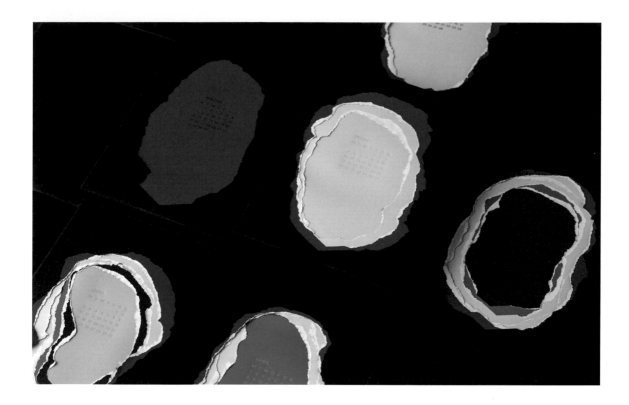

Dimensions:
DIN A5

Material(s):
Sirio Ultra Black of Fedrigoni Papers (115 gsm and 680 gsm),
Symbol Freelife Satin of Fedrigoni Papers (115 gsm)

Print & Finishing:
Hot-stamping

Different types of paper and paper of different weights were applied to the production of the calendar, transforming the abstract concept of time into a unique interactive experience of feeling and tearing. Due to the white picture-printing paper deliberately chosen, every action of tearing creates an edge in white, which contrasts well with the colours underneath.

Oryza Rice

Design: **Joanna Shuen**

Rice is a fundamental part of Japanese culture. Its cultivation and consumption have given fruition to customs, festivals and proverbs that shape Japan's identity. As such, the packaging design aims to capture these deep-rooted connections in the folding style, typography and calligraphy-style print pattern.

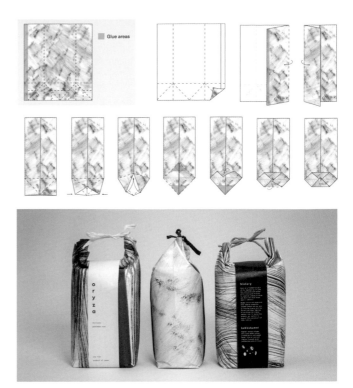

Dimensions:
100 mm × 70 mm × 220 mm

Material(s):
Paper

The brush and ink patterns allude to rice plants but are also informed by the specific use for the rice. For example, *hakumai* (white rice) is shown as a light pattern while *mochigome* (glutinous rice) features several dense and heavy strokes to reflect the pounding actions involved with preparing *mochi* (rice cake). Since sustainability was a key consideration, the rice is packaged in a sturdy paper bag that references historical examples of rice packaging. The choice of fold follows a traditional design—a paper pouch with a roll-top opening and paper rattan tie. When the contents are finished, the package can be reused as a gift bag.

Identité

Design Agency: **Seymourpowell**
Direction: **Mariel Brown, Nick Sandham, Neil Baron**
Design: **Robert Cooper**

Identité imagines how beauty products could be packaged with the application of artificial intelligence to be hyper-flexible and personalised. Identité explores the relationship between the power of algorithms to make decision for us and beauty products as a form of self-expression and personal identity.

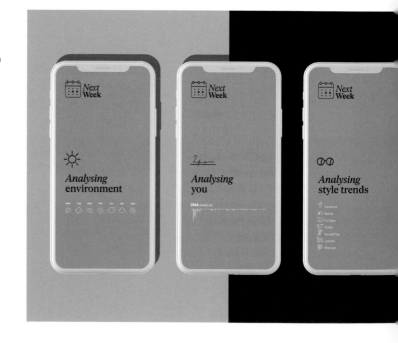

Dimensions:
85 mm × 28 mm × 150 mm

Material(s):
Formed paper, bio-degradable injection-moulded fibre

Print & Finishing:
Digital printing, hot-stamping with metallic foil

Identité's products are packed in perfectly portioned single use modules, made from formed paper and neatly collated into bio-degradable injection-moulded fibre boxes for regular delivery, creating fresher, more natural product formulas by reducing the need for preservatives. The result is an intelligent, sustainable beauty concept that harnesses artificial intelligence for the hyper-strategic beauty consumer of the future.

The Humble Co. Packaging Design

Design: **Fulya Kuzu**
Photography: **Fulya Kuzu**
Client: **The Humble Co.**

The Humble Co. develops reliable health and wellness products that are eco-friendly and socially responsible with an innovative twist. In terms of storytelling, the brand values a plain and concise language with the simple but efficient Scandinavian style that also gives an environmental-friendly touch.

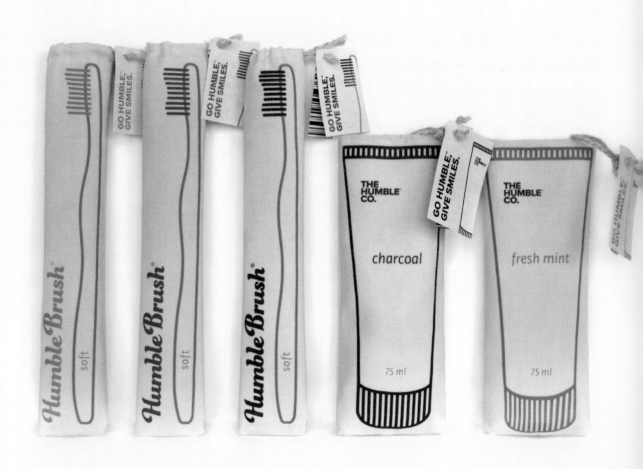

Dimensions:
Toothbrush: 40 mm × 10 mm × 210 mm
Toothpaste: 67.5 mm × 30 mm × 160 mm

Material(s):
Recycled polyester from PET bottles

Print & Finishing:
Silkscreen printing, sewing

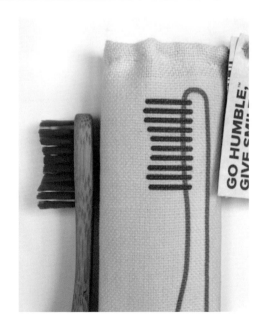

Therefore, this packaging chose eco-friendly recycled polyester from PET bottles, and the process reduced landfill, soil contamination and air and water pollution. The multifunctional packaging is waterproof, making it easy to store and transport, a good choice for travelling.

Immortal Fruit
Dried Hawthorn

Design: **Hsu Shih-Chien**
Supervision: **Hsu Shih-Chien**

As it is good for our health, hawthorn is also called 'immortal fruit' in Chinese culture, symbolising tenderness and protection. One set of the product includes hawthorn confections in three shapes—pellet, cake-shape and flower-shape, so Hsu Shih-Chien drew different illustrations for each type.

Dimensions:
200 mm × 85 mm

Material(s):
Baking paper, grey cloth, hemp rope

When choosing and testing the materials for the package, the designer turned to traditional sacks used to contain grains and dry foods for reference. He selected the grey cloth that is relatively soft and smooth and applied hemp rope to tie the bag with the aim to present the texture and touch commonly seen in farmers' daily life in East Asian culture.

LA MANUAL

Design: **Alba Morote, Nuria Torres, Paola Parodi**
Supervision: **Martin Azua, Eva Minguella**
Institution: **Elisava Barcelona School of Design and Engineering**

This project is a rebranding of La Manual Alpargatera, a business making espadrilles from 1940 based in Barcelona. Inspired by the elegance of Grace Kelly in a fisherman sweater wearing espadrilles, the designers wanted to add value to the shoes by designing a package to refine the product. By dipping the bag in indigo colour, they were trying to represent the Mediterranean sea and the fishermen in a poetic way.

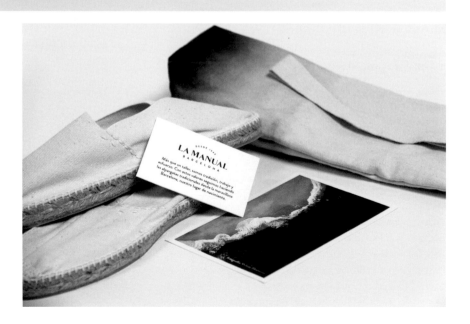

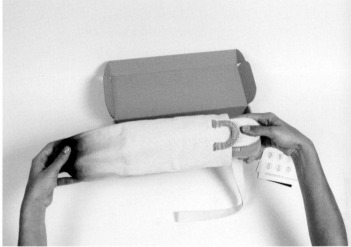

Dimensions:
350 mm × 150 mm × 60 mm

Material(s):
Organic cotton, esparto grass

Special Process:
Sewing, indigo dip dyeing

The designers created an organic cotton bag with a closure consisting of two straps—a long cotton one allowing the shoes to be carried and hung easily and a short one made of esparto grass helping to tie the bag with the long one. They chose cotton and esparto grass because these are the materials for making traditional espadrilles.

Tidy Products

Design Agency: **Unspoken Agreement**
Design: **Saxon Campbell, Don Campbell**

Tidy Products is a fictional antiseptic product brand designed for humans who desire a specific minimal aesthetic in their living spaces. For this personally branding and packaging project, Unspoken Agreement's main goal was to create an elevated and sophisticated visual aesthetic for a classic product.

Dimensions:
178 mm × 64 mm × 64 mm

Material(s):
Recycled plastic

Print & Finishing:
Plaster painting, speckled texturing

The design team achieved their goal by using the strong
juxtaposition between rough textures, which usually causes
scrapes or cuts, and clean and transparent liquid, which
typically cleans and heals those scrapes or cuts.

Elikto® Extra Virgin Olive Oil

Design Agency: **Chris Trivizas Design**
Design: **Chris Trivizas**
Photography: **Math Studio**
Client: **Kallistefi Olive Gems**

Elikto® is a superior quality extra virgin olive oil named after the title 'Dragon Eliktos' that was given to Hercules by the Orphics due to his versatility, similar to the versatility and persistence that embody the company's vision to track and offer the best extra virgin olive oil from selected olive groves all over Greece.

Dimensions:
159 mm × 83 mm (diametre)

Material(s):
Glass, wood

Print & Finishing:
Silkscreen printing

The packaging design was inspired by the marbles of the Panathenaic Stadium, the only stadium in the world made exclusively from Pentelic marble. For the promotion of the limited edition 100-millilitre bottle of Elikto®, a handmade box was created to imitate the skeleton boxes used to transport valuable archeological objects, aiming to protect their valuable content—the olive oil bottle.

G Candle Co. Packaging

Design Agency: **Prompt Design**

G Candle Company is a small-batch producer of scented candle collections with natural aroma and distinctive identity. The production process is meticulously controlled in every step. Each candle is hand-poured and crafted with charming design and attractive for those who prefer handmade merchandise.

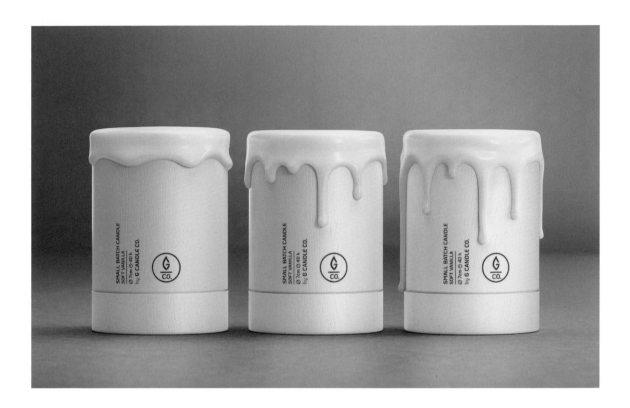

Dimensions:
140 mm × 90 mm (diametre)

Material(s):
Wood, wax

The wooden cylindrical package is selected to convey its natural product and crafted by dipping in sealing wax to create a unique wax flow appearance of candle teardrops around the top. This charming design can be enhanced with diverse mixing and changing colours of wax or wood cylinder. A wax seal is added on top to highlight its brand with the logo of a simple *G* whose figure is like a candle flame.

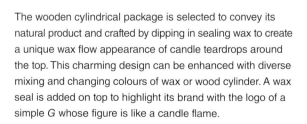

SENSE CUBE Packaging

Design: **Kim Hyun-Tae, Son Yoon-Joo**
Photography: **Kim Hyun-Tae, Son Yoon-Joo**

Compared to the past, people are
living in a limited tactile experience. In
response to these phenomena, the two
designers designed the SENSE CUBE to
remind people of their tactile experience.
SENSE CUBE's packaging is designed
to best protect each side of the cube
according to the core intent of the project.

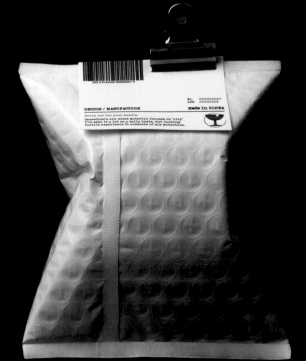

Dimensions:
60 mm × 130 mm × 170 mm

Material(s):
LDPE air cap

Typeface(s):
Corona 4 typewriter, Old typewriter, Tox typewriter

Print & Finishing:
Inkjet printing

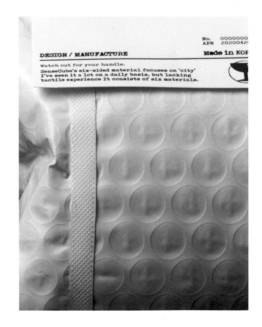

The air cap bag material, which is generally used to protect fragile items, is used. Through this packaging, the designers hope that SENSE CUBE's six tactile experiences will be fully delivered to consumers.

Frustum Package

Design Agency: **Masahiro Minami Design**
Art Direction: **Masahiro Minami**
Design: **Masahiro Minami**
Product Design: **Keita Hanazawa**

This is a prototype concept package design for the ceramic frustum cup designed by Keita Hanazawa. Since one of Mr Hanazawa's representative works is the paper mould lighting fixture, the designer used recyclable paper for the packaging design to extend his image.

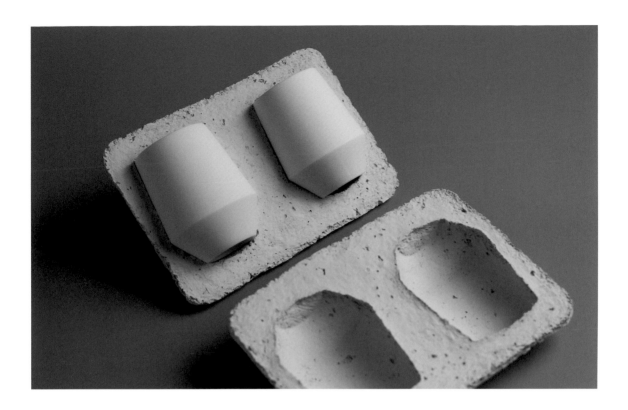

Dimensions:
205 mm × 145 mm × 70 mm

Material(s):
Recyclable paper

Two pieces of paper moulds in contrasting shapes cover and protect the ceramic inside; a sheet of paper with four square holes holds the paper moulds around; a paper band binds the different parts and makes the product easy to carry.

Palíndromo 5

Design Agency: **GRECO Design**
Photography: **Débora Colares**

Palíndromo (Palindrome) was created to publish
and publicise the portfolio of Rona Editora, a
traditional graphics and printing company. A flexible
and changeable format was conceptualised for the
editorial material with an interdisciplinary content
collaboration model proposed for each edition.
Different paper types and finishes were selected,
demonstrating a diversity of printing services and
a rich elaboration. Palíndromo, the name given to
the project, comes from Greek and means a word
that is the same written forwards or backwards.

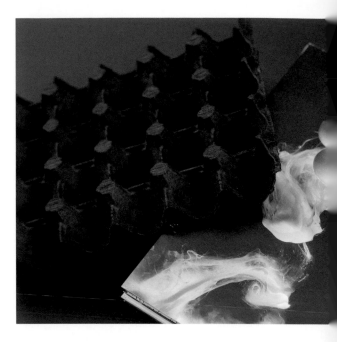

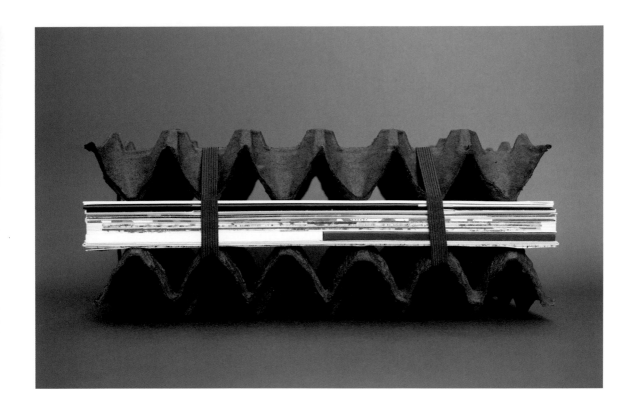

Dimensions:
210 mm × 290 mm × 70 mm

Material(s):
Egg carton, brassier-type buckles, elastic

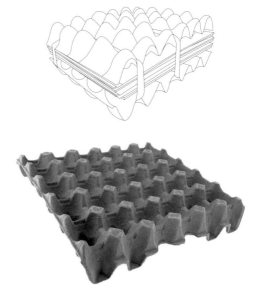

The theme of the third edition is 'Synthesis' expressed and illustrated by the Portuguese word 'OVO' (egg) and the egg carton. The booklets are in A4 format utilising only black and white, emphasising the synthesis process exercised by the employees.

Srisangdao Rice

Design Agency: **Prompt Design**

Thung Kula Ronghai has been a
well-known rice producing area of
Thailand, where its high quality of
rice production has been acclaimed
worldwide. Every year, the rice
production is limited in amount
in a controlled environment.
This organically produced rice,
therefore, is ensured of the best
quality and chemical-free and
thereby preserves the environment.
This is the origin of Srisangdao
Rice. Prompt Design's challenge
was to reflect all those organic rice
growing processes.

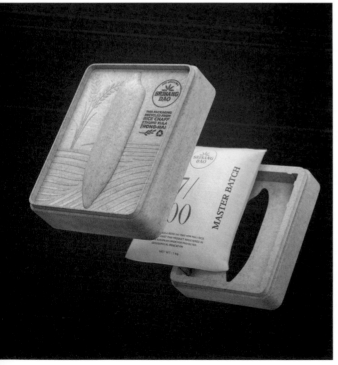

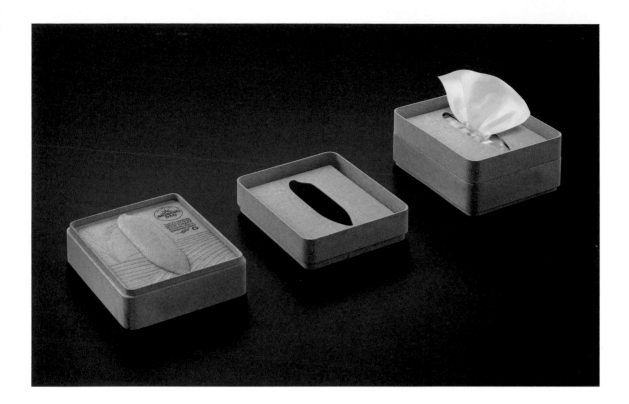

Dimensions:
160 mm × 215 mm × 80 mm

Material(s):
Rice chaff

Print & Finishing:
Embossing, hot-stamping

Prompt Design created the package by using the chaffs
which is the natural waste from the husking process. The
package is die-formed with the rice-shape embossing on
top of the box cover surrounded by the graphic lines and the
logo hot-stamped. Inside the box is filled with the rice sack
on which its lot number and other data are printed. Moreover,
this rice package may be later used as a tissue box.

Ayu Hiraki

Design Agency: **Masahiro Minami Design**
Art Direction: **Masahiro Minami**
Design: **Masahiro Minami**

In Japan, *himono* (dried fish) made from *ayu* (sweetfish) is a very rare food. The traditional way to make himono uses a bamboo colander to dry the fish. Inspired by the colander, Masahiro Minami renewed the package for Ayu Hiraki. Different from the round colander, the paper strips are knitted based on triangles to eliminate waste by avoiding the small holes in the corners of round shapes. Besides, the golden brand tag and a red pouch are inserted into the gap, making the package more elegant. The result is a better display and a more influential design with lower cost.

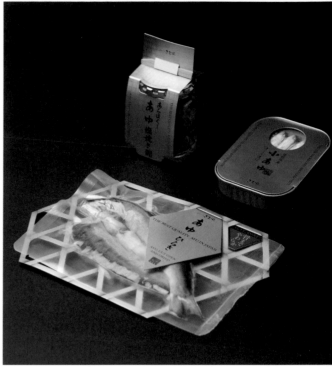

Dimensions:
222 mm × 142 mm × 5 mm

Material(s):
Paper

Print & Finishing:
Black and red ink on gold paper

Traditionally, himono is placed and dried on a bamboo colander, which gave Masahiro Minami the inspiration for this project. By making the paper strips flat and knitted, the designer brought freshness to the package while connecting it with himono producing tradition.

LA LUNE by ROCCA

Design Agency: **WWAVE DESIGN**
Design: **Amy Un Cho Ian, Ken Ho Ion Fat, Kenneth Ho**

The package was designed for the 2019 Mid Autumn Festival Gift box by a French pâtisserie based in Macau. The product is a set of two boxes of French pastry, chocolate and confectionery, created in a western artisanal method and recipe infused with Chinese flavour and elements.

Dimensions:
200 mm × 200 mm

Material(s):
Bamboo

Print & Finishing:
Embossing, hot-stamping

WWAVE DESIGN chose a hand-knitted bamboo container as an Oriental element, pairing with a contemporary and minimally designed paper sleeve of a different texture to create a modern style for this traditional Chinese festival.

SONAE UME

Design Agency: **KITADA DESIGN Inc.**
Art Direction: **Shingo Kitada**
Production: **Jumpei Takeuchi**
Client: **BambooCut Inc.**

SONAE UME is a brand of Japanese *umeboshi* (sour plum) made by the craftsperson Sachiko Norimatsu. After the earthquake in Kumamoto, Japan, in 2016, SONAE UME was proposed and produced as a type of food for emergencies. Although it has developed beyond emergency food, its package should be small and easy to carry.

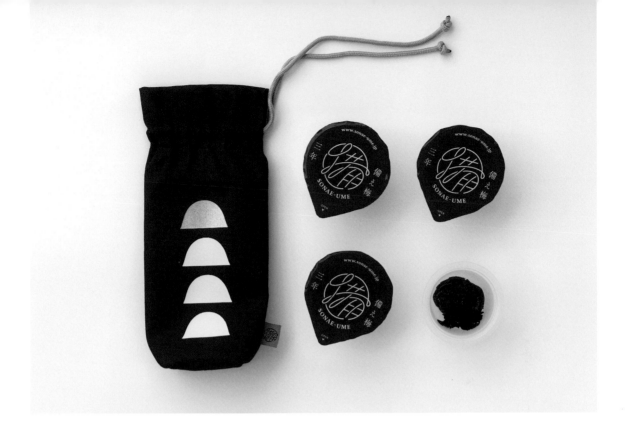

Dimensions:
190 mm × 55 mm × 55 mm

Material(s):
Cotton, PET

Print & Finishing:
Silkscreen printing

The design team chose a special drawstring bag to contain four plums each sealed in small plastic cups, making the package look like an amulet that carries a meaning of wishes and good luck.

CHI-SHANG
Valley Rice

Design: **Hsu Shih-Chien**
Supervision: **Lin Ting-Chien**

The keys to CHI-SHANG Valley Rice are the clear spring water, the unique climate, the fertile soil of the alluvial plain and the quality species of the Taikeng No. 2 rice. Hsu Shih-Chien represented these four concepts with totems in a simple and fresh visual style.

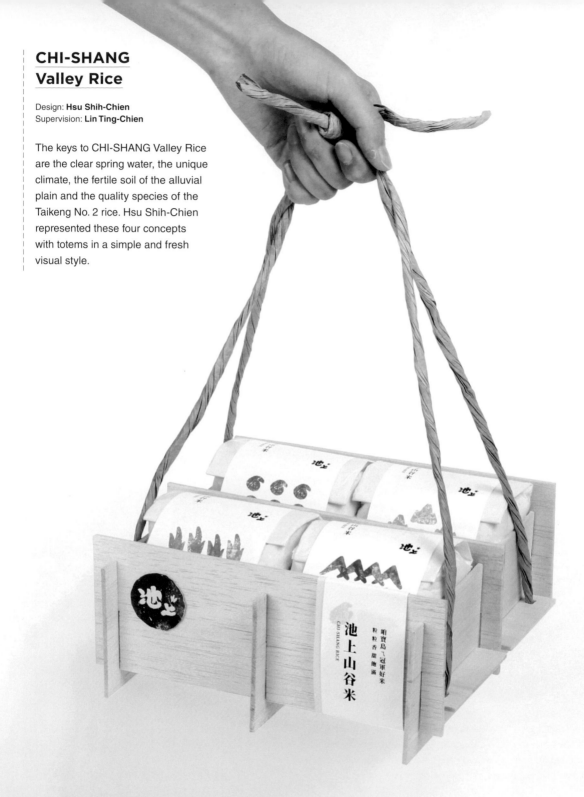

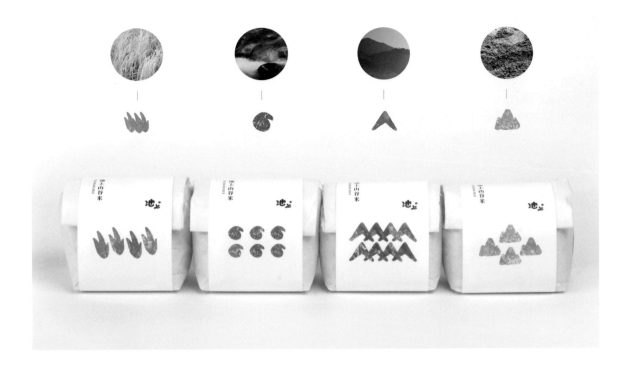

Dimensions:
Outer carrier: 250 mm × 280 mm × 75 mm
Rice pack: 95 mm × 95 mm × 95 mm

Material(s):
Wooden board, double-layered paper,
coarse paper string

The outer package is made of wooden boards built into the shape of the Chinese character '田' (soil, field). The inner bags utilise double-layered cotton paper to wrap the rice, which look full, round, cute and friendly like the rice itself.

Supha Bee Farm Honey

Design Agency: **Prompt Design**

In Thailand, the honey product market is competitive, filled with various package designs. Thus, Prompt Design had to develop a brand identity for Supha Bee Farm which should be able to deliver its specific advantage. Supha Bee Farm is one of the two main honey producers in Thailand which has its own bee farms with rearing and breeding bees facilities, producing an outstanding product of 100% real pure honey.

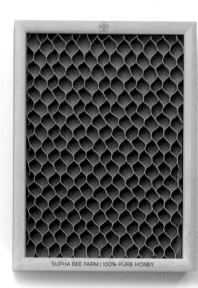

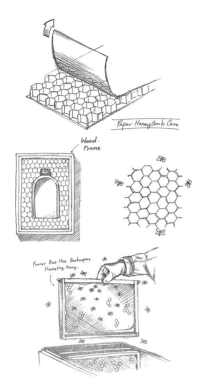

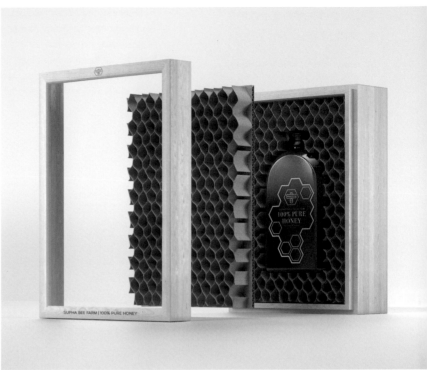

Dimensions:
185 mm × 240 mm × 90 mm

Material(s):
Wood, honeycomb core paper

The package is specially designed and inspired by the structure of a beehive frame. The paper honeycomb is used together with the wooden box to emphasise the feeling that the honey bottle inside is virtually from the beehives. The logo of 'SB', the abbreviation of the brand, was designed simply to resemble a bee.

Irati Patxarana - Artisans in 2019

Design Agency: **Bonita Estudio**
Photography: **Nere Oria**

Irati constitutes a complete graphic language that combines typography and corporate identity. 'Irati Regular' is a typeface which reflects the Basque-Navarrese culture in a subtle and delicate way. It is inspired by the unspoilt landscape of the Irati Forest, with the aim of giving personality to this new sloe gin (patxaran).

Dimensions:
67 mm × 95 mm

Material(s):
Tintoretto gesso paper (120 gsm), 100% cotton

Typeface(s):
Irati Regular

Print & Finishing:
Offset printing, serigraphy

'Irati Patxarana - Artisans in 2019' opts for the selection of raw materials and hand-crafted finishes, based on a colour gamut that mixes the green tones of nature with the deep red colour of this Navarrese drink.

Jo-Chu

Design Agency: **NOSIGNER**
Art Direction: **Eisuke Tachikawa**
Graphic Design: **Eisuke Tachikawa,
Ryota Mizusako, Jin Nagao**
Product Design: **Eisuke Tachikawa,
Daichi Komatsu**
Client: **NAORAI CO., Ltd**

Sake tends to change in taste over time, making the quality difficult to control. To convey the true taste and appeal of sake, NOSIGNER worked on the overall branding and design of Jo-chu, a new sake developed by NAORAI CO., Ltd. Jo-chu is made by a new method of low-temperature distillation that does not compromise the unique flavour and taste of sake. Moreover, it is possible to prevent quality deterioration.

Dimensions:
Bottle: 227 mm × 109 mm (diametre)
Wooden box: 270 mm × 126 mm × 124 mm

Material(s):
Glass, paper, Paulownia wood

Print & Finishing:
Hot-stamping, silkscreen printing

The shape of the bottle is inspired by the shape of a balloon filled with water, recreating the natural shape formed by the tension of the liquid. Besides, the distinctive design of *shide* (a zigzag-shaped paper streamer used in the Japanese religion *Shinto*) is added onto the bottle.

Costenc

Design Agency: **Zoo Studio**
Art Direction: **Gerard Calm, Xevi Castells**
Design: **Xevi Castells**

Set & Ros promotes a new project of single vineyard wine from the variety Malvasia from Sitges, Spain, a very special autochthonous variety that is in process of recovery. Set & Ros asked Zoo Studio for the naming and the design of the bottle. Costenc is a wine with characters—the result of vineyards that have suffered the influence of the sea, the salinity and sandy soil.

Dimensions:
290 mm × 85 mm (diametre)

Material(s):
Eroded glass and paper, wax

The design studio wanted to transfer the influence of the sea to the design of the bottle thus represented the erosion caused by the waves and sea sand both on the label and bottle. The design of the label is sober and elegant, showing only the name and description that refers to the proximity to the sea. The turquoise wax cap unrolls this relationship with the coast.

Holiday Beer series

Design Agency: **Polygraphe Studio**
Art Direction: **Yann Carrière**
Creative Direction: **Sébastien Bisson**
Design: **Yann Carrière**

The package was created for a beer
family produced in a small batch that
has been distributed to the clients and
collaborators to celebrate the end of
the year. Polygraphe Studio worked on
the theme of snow and frost, a parallel
between the local climate during
wintertime and the characteristic of
these easy-drinking light beers, best
appreciated cold.

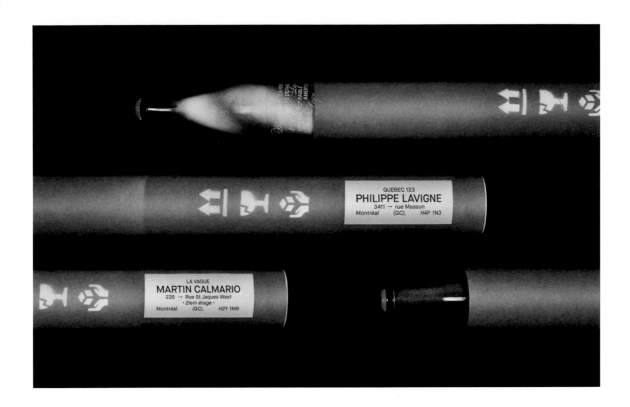

Dimensions:
1219 mm × 76 mm (diametre)

Material(s):
Glass, cardboard

Print & Finishing:
Spray painting

To simulate the effect of snow and frost, they manually
spray-painted each bottle in order to create a unique object.
It is then inserted into a personalised shipping tube.

Enrejado

Design Agency: **F33**
Client: **Enrejado**

Enrejado is a new premium gin of two versions—Dry Gin in black and Red Gin in white.

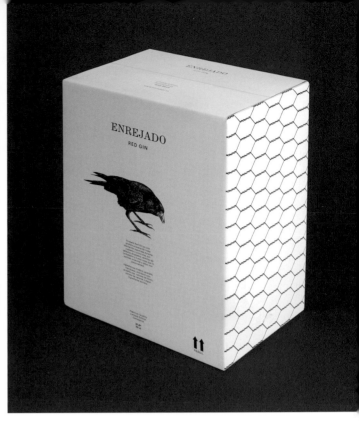

Material(s):
Glass, aluminium

Print & Finishing:
Silkscreen printing on tinted glass

To design a special package for it, F33 created a lacquered metal mesh, using aluminium, in a traditional way one by one. Complemented by the illustrations of birds, the entire package looks like a mesh trapping the bird, which further illustrates the brand's name 'Enrejado' (trellis, lattice).

Dinamite

Design Agency: **327 Creative Studio**

Dinamite is a unique sparkling wine collection from the clay-limestone soils of Bairrada, Portugal. With a distinctive and bold identity, the concept of this three-bottle edition comes from its own name—Dinamite (dynamite). The identity developed for this project arose from the need to differentiate and number the three wines. Accordingly, 327 Creative Studio created three geometric sets—based on the dynamite silhouette—that became the face of this collection.

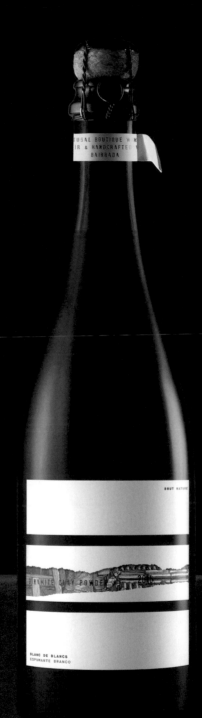

Dimensions:
310 mm × 80 mm (diametre)

Material(s):
Glass bottle, paper, foil

Print & Finishing:
Offset printing, hot-stamping

With the intention of creating a label that would allow the consumers to interact with the product and express the metaphorical representation of 'explosion', the design studio used an overlay label with a 'fuse' that can be pulled and ripped, revealing the name of the product.

Oblivion

Design Agency: **Artware**
Client: **Oblivion**

As the most valuable fruit of the Mediterranean, the olive has defined the history, society and daily life of the people in that region over the centuries. Oblivion, an organic and ultra-premium series of extra virgin olive oil, had been seeking its identity. Artware tried to retrieve those 'forgotten' characteristics of Oblivion's substance and quality, which are directly related to its production, history and essence.

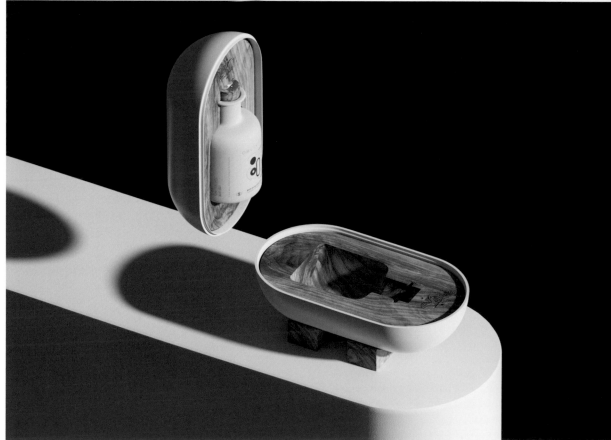

Dimensions:
236 mm × 128 mm × 134 mm

Material(s):
Wood, ceramic

Print & Finishing:
Ceramic coating, silkscreen printing

By applying wood and ceramic to the outer shell and
the bottle hidden inside, Artware not only designed a
package matching Oblivion's value and philosophy but also
attempted to remind the consumers of the origin of olives.

PACKAGING WITH IMPRESSIVE VISUALS

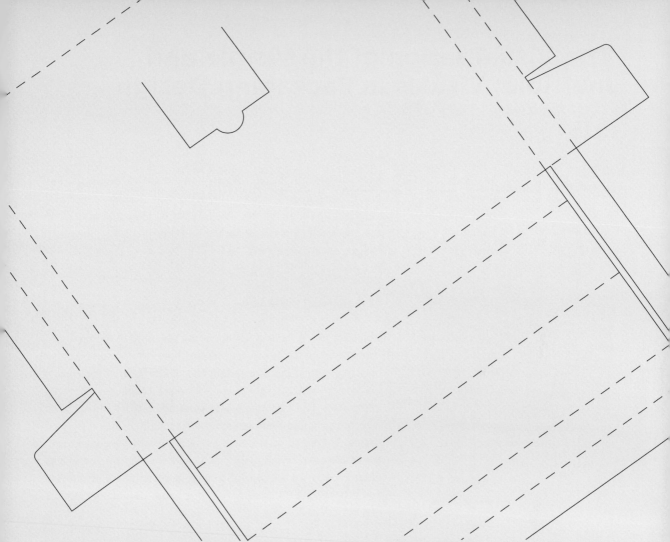

While shopping, the first thing that catches consumers' eyes must be the visual effect of a product. Shape, colour, typography and various graphic elements of packaging design collaborate to visualise the product's nature and value, stimulating consumers' curiosity as well as communicating the key information of the products. This section introduces how visual language carries information in packaging design and displays impressive packages that consumers can hardly take their eyes off.

The Visualisation of the Visible and Invisible: Visuals in Packaging Design

Like any other languages in the world, visual language in packaging design carries enormous messages that directly strike our eyes before our brains start consciously processing them. The following paragraphs discuss how visual language is used to visualise 'the visible'—the practical information about the products—and 'the invisible'—the brand's culture, values and philosophy behind.

Packages of lipsticks designed by Transwhite Studio for the brand 3rd Universe.

Colours and illustrations that Kingdom & Sparrow employed to differentiate the skincare products in a collection of Scence.

1. Visualising the Visible:

Referring to the information about the contents of the products, 'the visible' explains their background, property, features and so on, ranging from the flavours of soda, the colours of lipsticks, to the origins of wines, the genres of music. Of course, all those information can be written on the labels, yet we are humans who have limited time in their life and limited space in their brains. Therefore, in terms of helping consumers dealing with the information on an overloaded market stuffed by countless products, the significance of visuals in packaging is to simplify that information and shorten their time spent on searching for the needed (or unneeded).

Colours and illustrations are some of the effective visual elements in visualising the visible. Take the packaging of cosmetics as an example: the according colours are usually printed on the packages of lipsticks, blushers and so on so that the consumers can easily recognise which match their expectations the most before further tries. Illustrations are sometimes used for indicating the ingredients or origins of the products. For instance, lip balms of peppermint flavour are illustrated with leaves of mint more often than not, the ones of honey flavour with bees or honeycombs. On the labels of wines, there are sometimes illustrations of the breweries or grapes plantations, showing the producing places of the wines. Besides, applying colours and illustrations to product lines or collections is also a helpful approach to distinguishing different products within the lines or collections.

Red envelope designed by Wing Yang for the 2020 Chinese New Year.

Packages centred on typography designed by moodley design group for the brand Tieschen.

2. Visualising the Invisible:

The visualisation of the invisible lies in creating the vibes, communicating the culture and conveying the brand's values and philosophy, of which the significance is to build the connection between the brands and the audience, to reinforce the sustainability of the brands and strengthen their audience's attachment to them.

Colour palette plays an important role in this practice. When deciding the colour palette, there could be three ways to get started. First, making use of colour psychology is a direct and quick method to set the tone of the products. For example, red is related to passion and energy, blue to peace and quietness. Second, utilising colours with cultural meaning is useful in attracting the audience under a certain cultural background. To be specific, red means good luck and happiness in Chinese culture. Last but not least, playing with the brand's intrinsic colours helps to maintain consistency and to harmonise the new and existing products under one brand.

Other elements such as typography, logos, print and finishing in packaging might also have a great impact on shaping the brand. Bold typography and logos assist the brand to deepen its impression on the audience and to make it stand out on the shelves. Efforts on printing and finishing are especially helpful for packages of limited editions or gift sets. For example, embossing would add textures to the packages, and hot-stamping with metallic foils would upgrade the brand.

Being universally recognisable, visual language is beyond cultures and languages. Thus, in this globalising world, it is an effective tool to communicate the visible and invisible information to consumers from all over the world and to lend the brands a hand to win a place on the international market.

ENDLESS GEOMETRIC ART

Design Agency: **RHYTHM INC.**
Design: **Junichi Hakoyama**

Through the black-and-white cubes with parts of a circle, the collective entity brings various expressions depending on how they are put together. The collective graphics that are focused on continuity have infinite possibilities of expressions.

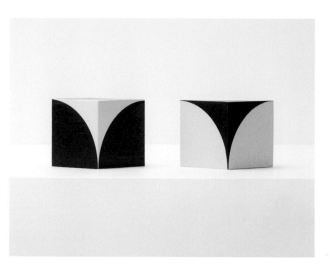

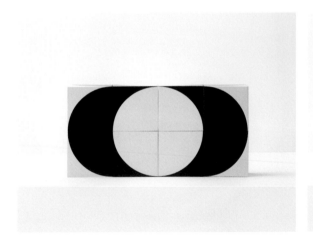

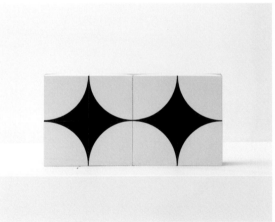

Dimensions:
75 mm × 75 mm × 75 mm

Material(s):
Paperboard

Print & Finishing:
Single-colour printing

● ○
C: 0   C: 0
M: 0   M: 0
Y: 0   Y: 0
K: 100 K: 0

Spectrum

Design: **Olga Ivanova**
Supervision: **Pavel Borisovskiy**

Spectrum is a brand producing tinted contact lenses for daily use that are presented in 16 tints. The brand's name reflects the features of the product—vibrancy, broad variety and multitude of colours. The packaging design considered precision, comfort and vibrancy. Medical pictograms related to optics comprise its foundation. Besides, the designer's original intention was to create an image of 'the outer space in a box'. To achieve that, gradients resembling small planets were added to the style, and the general stylised design was transformed into technical records and symbols. All style objects are distributed across a grid that can change their shapes depending on the medium used.

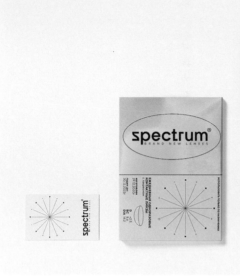

Dimensions:
A: 300 mm × 240 mm × 50 mm
B: 150 mm × 50 mm × 30 mm
C: 120 mm × 90 mm × 20 mm

Material(s):
**Plastic, cardboard, paper,
aluminium**

Typeface(s):
Aventa Variable, TT Norms

Print & Finishing:
Flexography, digital printing

C: 48	C: 36	C: 1
M: 91	M: 80	M: 51
Y: 0	Y: 0	Y: 0
K: 0	K: 0	K: 0

C: 52	C: 4	C: 22
M: 0	M: 0	M: 0
Y: 58	Y: 96	Y: 71
K: 0	K: 0	K: 0

C: 36	C: 1	C: 88
M: 80	M: 51	M: 70
Y: 0	Y: 0	Y: 0
K: 0	K: 0	K: 0

C: 1	C: 1	C: 0
M: 16	M: 51	M: 67
Y: 80	Y: 0	Y: 76
K: 0	K: 0	K: 0

C: 90	C: 0	C: 75
M: 82	M: 86	M: 32
Y: 0	Y: 3	Y: 0
K: 0	K: 0	K: 0

C: 26	C: 25	C: 2
M: 0	M: 27	M: 17
Y: 7	Y: 0	Y: 0
K: 0	K: 0	K: 0

C: 61	C: 52	C: 4
M: 46	M: 0	M: 0
Y: 0	Y: 58	Y: 96
K: 0	K: 0	K: 0

C: 4	C: 0	C: 36
M: 0	M: 67	M: 80
Y: 96	Y: 76	Y: 0
K: 0	K: 0	K: 0

C: 0	C: 85	C: 62
M: 96	M: 78	M: 0
Y: 78	Y: 0	Y: 45
K: 0	K: 0	K: 0

C: 60	C: 85	C: 5
M: 0	M: 78	M: 93
Y: 80	Y: 0	Y: 0
K: 0	K: 0	K: 0

C: 52	C: 1	C: 1
M: 0	M: 16	M: 51
Y: 58	Y: 80	Y: 0
K: 0	K: 0	K: 0

Laroché Chocolate

Design: **Martin Naumann, Andrius Martinaitis**
Photography: **Mantas Zinkevičius**

Martin Naumann and Andrius Martinaitis created a vibrant packaging concept for their new chocolate creations for Laroché Confectionery from Lithuania. The abstract textures were parametrically generated to visualise the unique taste experience of each flavour. The series contains packages for the flavours ruby, caramel, bittersweet and milk chocolate.

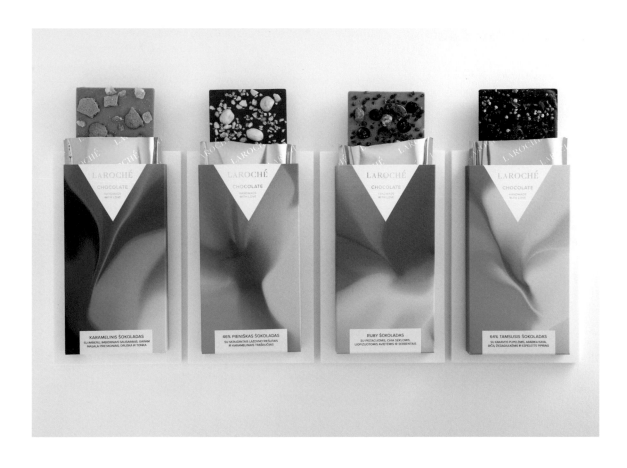

Dimensions:
170 mm × 79 mm × 10 mm

Material(s):
Paper (305 gsm), plastic bag

Typeface(s):
Custom type

Print & Finishing:
Hot-stamping with soft-touch foil and gold foil

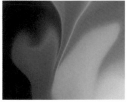

C: 10   C: 80
M: 80   M: 90
Y: 0   Y: 0
K: 0   K: 0

C: 0   C: 0
M: 35   M: 85
Y: 95   Y: 85
K: 0   K: 0

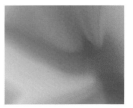

C: 0   C: 0
M: 60   M: 95
Y: 100   Y: 10
K: 0   K: 0

C: 75   C: 10
M: 10   M: 25
Y: 10   Y: 15
K: 0   K: 0

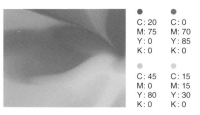

C: 20   C: 0
M: 75   M: 70
Y: 0   Y: 85
K: 0   K: 0

C: 45   C: 15
M: 0   M: 15
Y: 80   Y: 30
K: 0   K: 0

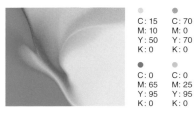

C: 15   C: 70
M: 10   M: 0
Y: 50   Y: 70
K: 0   K: 0

C: 0   C: 0
M: 65   M: 25
Y: 95   Y: 95
K: 0   K: 0

Red Envelope 2020

Design: **Wing Yang**
Copywriting: **Longlong**

Red Envelope 2020 is a combination of
traditional Chinese style and concise
geometric style, transforming traditional
images of Spring Festival such as lanterns,
fireworks and rats (representing Year of the
Rat) into geometric shapes. The red and
golden colour palette is reminiscent of a
typical Spring Festival vibe.

Dimensions:
120 mm × 190 mm

Material(s):
Linen paper

Typeface(s):
Source Han Serif

Print & Finishing:
UV varnishing

● C: 10
M: 93
Y: 85
K: 0

● C: 2
M: 21
Y: 54
K: 0

○ C: 0
M: 0
Y: 0
K: 0

Share the Joy of Floral

Design Agency: **studioWMW**
Design Direction: **Sunny Wong**
Design: **Kylie Lee**
Client: **Polytrade Paper Corporation Ltd.**

This project started from studioWMW's exploration of an innovative and modern aesthetic of red packets beyond the tradition of reddish colors and fancy patterns. They played with the meaning of various flowers representing different blessings. Further, the concept became 'flower in the vase' with a sleeve as the vase. Equipped with a stand at the back, each red packet can be transformed into a decorative table standee. The packets are made of quality fancy paper from Polytrade Paper and have employed various printing techniques such as laser cutting and 3D embossing. Through those packets, studioWMW added a layer of modernity to the traditional value of the Spring Festival by making the red packets modern collectibles.

Dimensions:
95 mm × 175 mm × 45 mm

Material(s):
Paper, plastic

Typeface(s):
Miller Display, Circular Std

Print & Finishing:
Die-cutting, hot-stamping with matte gold and matte silver foil, 3D embossing

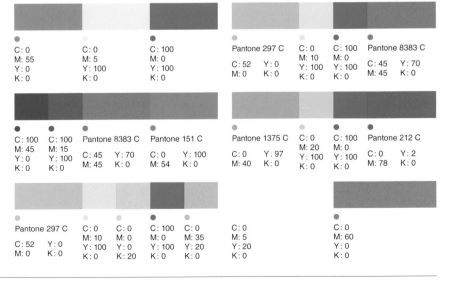

C: 0	C: 0	C: 100
M: 55	M: 5	M: 0
Y: 0	Y: 100	Y: 100
K: 0	K: 0	K: 0

Pantone 297 C	C: 0	C: 100	Pantone 8383 C
C: 52 Y: 0	M: 10	M: 0	C: 45 Y: 70
M: 0 K: 0	Y: 100	Y: 100	M: 45 K: 0
	K: 0	K: 0	

C: 100	C: 100	Pantone 8383 C	Pantone 151 C
M: 45	M: 15	C: 45 Y: 70	C: 0 Y: 100
Y: 0	Y: 100	M: 45 K: 0	M: 54 K: 0
K: 0	K: 0		

Pantone 1375 C	C: 0	C: 100	Pantone 212 C
C: 0 Y: 97	M: 20	M: 0	C: 0 Y: 2
M: 40 K: 0	Y: 100	Y: 100	M: 78 K: 0
	K: 0	K: 0	

Pantone 297 C	C: 0	C: 0	C: 100	C: 0	C: 0
C: 52 Y: 0	M: 10	M: 0	M: 0	M: 35	M: 5
M: 0 K: 0	Y: 100	Y: 0	Y: 100	Y: 20	Y: 20
	K: 0	K: 20	K: 0	K: 0	K: 0

C: 0
M: 60
Y: 0
K: 0

NORD STREAM

Design Agency: **LOCO Studio**
Design: **Artem Petrovsky,
Evgeniya Petrovskaya, Andrey Shkola,
Vladimir Zotov**

NORD STREAM is a product line of preserved foods, whose main competitive advantages are interactivity and responsiveness. The brand consists of five different tastes: sardine, smoked mussels, crab, octopus and squid. Each package is distinguished from the others due to its graphic and colour identification. While creating the concept, LOCO Studio came up with the idea that communication of the product with its potential customers should be done with the help of the hidden animation, which is based on the work of the moiré pattern. The picture on the top starts moving when you open the package. It impresses and involves customers into the process of communicating with the product.

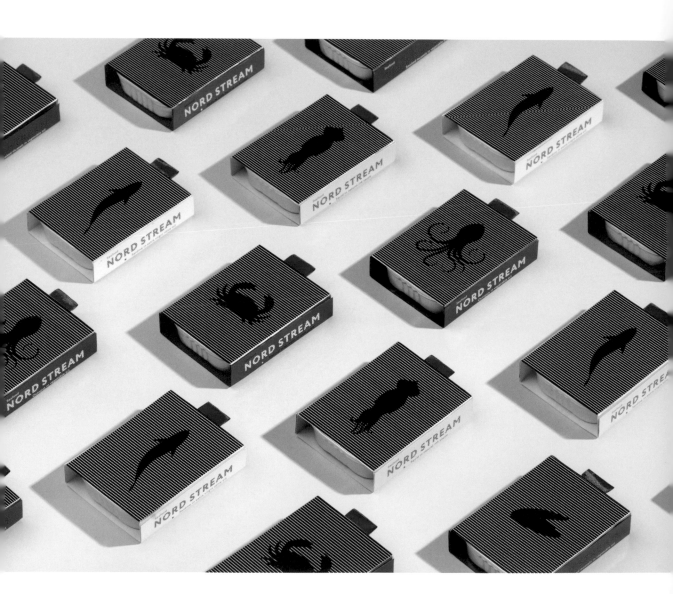

Dimensions:
78 mm × 108 mm

Material(s):
**Plastic (3 mm thick),
paper (300 gsm)**

Typeface(s):
Circe

Print & Finishing:
Offset printing

○	●
Pantone warm gray 1 C	Pantone 539 C
C: 10 Y: 11	C: 100 Y: 0
M: 10 K: 0	M: 43 K: 83

●	●
Pantone 539 C	Pantone 199 C
C: 100 Y: 0	C: 0 Y: 79
M: 43 K: 83	M: 100 K: 0

●	○
Pantone 199 C	Pantone warm gray 1 C
C: 0 Y: 79	C: 10 Y: 11
M: 100 K: 0	M: 10 K: 0

○	●
Pantone 2905 C	Pantone 199 C
C: 43 Y: 0	C: 0 Y: 79
M: 3 K: 0	M: 100 K: 0

●	○
Pantone black 3 C	Pantone warm gray 1 C
C: 74 Y: 71	C: 10 Y: 11
M: 52 K: 90	M: 10 K: 0

BEAN TO BAR

Design Agency: **Zoo Studio**
Art Direction: **Gerard Calm, Gisela Sole**
Design: **Gisela Sole**

The commission was to create an avant-garde design that explains and gives value to the original process of making chocolate, the origin and the raw material with the concept 'Bean to Bar'. Zoo Studio chose a unique design— an 'object-pack'. The booklets are manually sewn with a thread holding the booklet to the chocolate bar. The studio used the four sheets of the book to graphically communicate the chocolate's origin and characteristics and the manufacturing process. The interior of the booklets was designed with a single ink, a very simple language with few resources. Small illustrations and symbols of naive aesthetics and reminiscent of ethnic and tribal arts were also created.

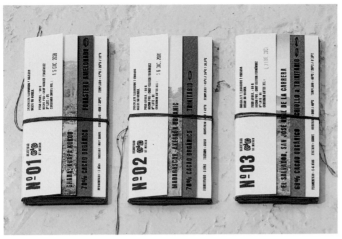

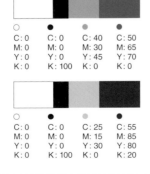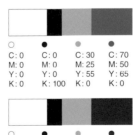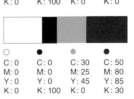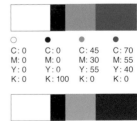

Dimensions:
140 mm × 65 mm

Material(s):
**Paper, aluminium paper,
cotton thread**

○	●	●	●
C: 0	C: 0	C: 40	C: 50
M: 0	M: 0	M: 30	M: 65
Y: 0	Y: 0	Y: 45	Y: 70
K: 0	K: 100	K: 0	K: 0

○	●	●	●
C: 0	C: 0	C: 30	C: 70
M: 0	M: 0	M: 25	M: 50
Y: 0	Y: 0	Y: 55	Y: 65
K: 0	K: 100	K: 0	K: 0

○	●	●	●
C: 0	C: 0	C: 45	C: 70
M: 0	M: 0	M: 30	M: 55
Y: 0	Y: 0	Y: 55	Y: 40
K: 0	K: 100	K: 0	K: 0

○	●	●	●
C: 0	C: 0	C: 25	C: 55
M: 0	M: 0	M: 15	M: 85
Y: 0	Y: 0	Y: 30	Y: 80
K: 0	K: 100	K: 0	K: 20

○	●	●	●
C: 0	C: 0	C: 30	C: 50
M: 0	M: 0	M: 25	M: 80
Y: 0	Y: 0	Y: 45	Y: 85
K: 0	K: 100	K: 0	K: 30

○	●	●	●
C: 0	C: 0	C: 40	C: 60
M: 0	M: 0	M: 25	M: 80
Y: 0	Y: 0	Y: 30	Y: 60
K: 0	K: 100	K: 0	K: 25

Holiments

Design Agency: **Parámetro Studio**
Illustration: **Kitty Ramos**

Holiments is a brand for highly nutritional chocolate products made from the best quality and natural ingredients which are gluten-free, dairy-free and keto-friendly. Parámetro Studio wanted to develop a real, natural and honest brand. The name comes from two words—'holy' and 'aliments'—directly referring to the preparation of the products and the hands behind. They decided to use illustrations as a key element in the design so that the ingredients in each product can be seen clear and loud. Holiments uses warm colours to create this nostalgia of looking throughout old recipe books but with a slight pop of colours to make it stand out on the shelf.

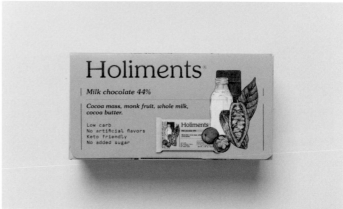

Dimensions:
148 mm × 138 mm

Material(s):
BOPP (biaxially-oriented polypropylene)

Typeface(s):
**ITC Souvenir Std Light Italic,
ITC Souvenir Std Medium Italic,
Apercu Mono Regular,
Rekey Regular**

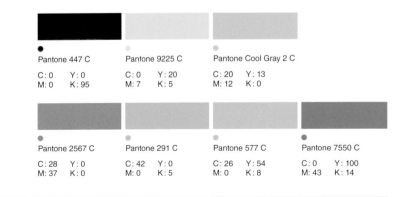

Pantone 447 C

C: 0   Y: 0
M: 0   K: 95

Pantone 9225 C

C: 0   Y: 20
M: 7   K: 5

Pantone Cool Gray 2 C

C: 20   Y: 13
M: 12   K: 0

Pantone 2567 C

C: 28   Y: 0
M: 37   K: 0

Pantone 291 C

C: 42   Y: 0
M: 0   K: 5

Pantone 577 C

C: 26   Y: 54
M: 0   K: 8

Pantone 7550 C

C: 0   Y: 100
M: 43   K: 14

Phú Quốc Sea Gift

Design Agency: **GM Creative**
Design: **Kien Kit**
Photography: **Oanh Pham, Tran Le**
Client: **Phú Quốc Sea Gift**

Phú Quốc Sea Gift was born as the unique combination of high-quality seafood and the finest natural ingredients of Phú Quốc Island such as pepper, fish sauce and so on. Every day, the skillful staff follow strict recipes that use traditional methods combined with Japanese technology to create delicious and nutritious seafood snacks.

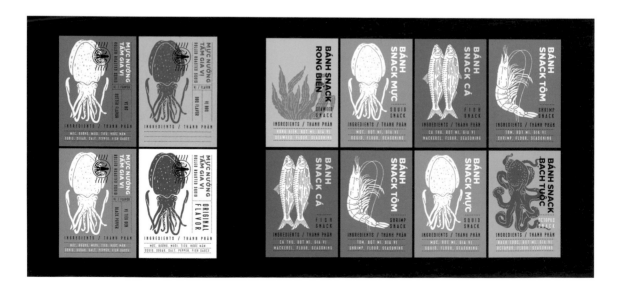

Typeface(s):
UTM Pacific Standard-Regular

○ C: 0 M: 0 Y: 0 K: 0
● C: 0 M: 75 Y: 84 K: 0
● C: 100 M: 97 Y: 29 K: 20

● C: 0 M: 25 Y: 60 K: 0
● C: 0 M: 75 Y: 84 K: 0
● C: 100 M: 97 Y: 29 K: 20

● C: 45 M: 55 Y: 20 K: 0
○ C: 0 M: 0 Y: 0 K: 0
● C: 100 M: 97 Y: 29 K: 20

● C: 70 M: 50 Y: 20 K: 0
○ C: 0 M: 0 Y: 0 K: 0
● C: 5 M: 10 Y: 45 K: 0

● C: 55 M: 20 Y: 50 K: 0
○ C: 0 M: 0 Y: 0 K: 0
● C: 100 M: 97 Y: 29 K: 20

● C: 40 M: 5 Y: 80 K: 0
● C: 60 M: 30 Y: 90 K: 0
● C: 100 M: 97 Y: 29 K: 20

DIVINE 3

Design: **Karla Heredia Martínez**

DIVINE 3 is an organic brand with love and respect for natural values. The representation of the brand should include those elements in their personality and overall style. To achieve that, the designer gave the brand a fresh and sincere look combined with abstract illustrations about natural beauty—landscapes, plants, mountains, water, earth and so on—with an exaggerated and energetic colour palette.

Material(s):
Organic cotton paper

Typeface(s):
Optima, handmade type

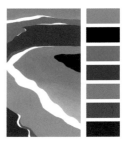

C: 52   C: 81   C: 93   C: 61   C: 12
M: 18   M: 66   M: 48   M: 10   M: 1
Y: 7   Y: 56   Y: 91   Y: 98   Y: 93
K: 0   K: 69   K: 63   K: 0   K: 0

C: 89   C: 85   C: 62   C: 66   C: 63
M: 44   M: 58   M: 0   M: 27   M: 7
Y: 93   Y: 69   Y: 67   Y: 10   Y: 42
K: 53   K: 79   K: 0   K: 1   K: 0

C: 64   C: 97   C: 3   C: 10
M: 22   M: 87   M: 64   M: 90
Y: 8   Y: 47   Y: 87   Y: 90
K: 0   K: 64   K: 0   K: 2

C: 10   C: 13   C: 12   C: 39
M: 90   M: 25   M: 41   M: 75
Y: 92   Y: 59   Y: 89   Y: 66
K: 2   K: 2   K: 2   K: 58

C: 56   C: 12   C: 17   C: 24
M: 9   M: 1   M: 8   M: 34
Y: 99   Y: 93   Y: 53   Y: 36
K: 0   K: 0   K: 0   K: 9

C: 29   C: 32   C: 36
M: 45   M: 64   M: 88
Y: 29   Y: 45   Y: 82
K: 8   K: 29   K: 55

Modesta Cassinello.
Solid Hair Essence

Design Agency: **Plácida**
Photography: **Cris Beltrán**
Client: **Modesta Cassinello**

Modesta Cassinello is Mediterranean cosmetics inspired by the ingredients and habits of this culture. Its hair line has included three new products—H05 shampoo with grapefruit essence, H06 conditioner with almond oil and H07 shampoo with sea salt. The graphic image is resolved with the use of a reduced black and white palette and a classic serif typography that give the cosmetic line its own personality and unisex language.

Modesta
Cassinello

H05

Champú Sólido
Solid Shampoo

Citrus paradisi

Cosmética mediterránea con aloe vera
y esencia de pomelo
Mediterranean cosmetics with aloe vera
and grapefruit essence

GENDERLESS / FOR ALL KIND OF HAIR

Esp /
Avanza y elige hoy para el cuidado de tu cabello la opción más
efectiva y la más respetuosa con nuestro medio ambiente por
su tamaño reducido, cero plásticos y activos de origen natural
que garantizan un cabello limpio, hidratado y revitalizado.

¿Cómo utilizarlo? Frota el champú con las manos hasta obte-
ner una fina espuma y deposítala sobre la raíz del cabello pre-
viamente mojado. Masajea suavemente, disfrutando el mo-
mento y enjuaga bien. Repite la aplicación y continúa con
nuestro acondicionador sólido H06.

Eng /
Go ahead and choose today the most effective option for your
hair, and also the most respectful with our environment because of its reduced size, zero plastics, and natural origin actives that offer a clean, hydrated and revitalized hair.

How to use it? Rub the shampoo bar between your hands un-
til a fine lather is obtained. Massage gently, enjoying the moment, and deposit on your hair roots, pre-
viously wet. Repeat and continue with our solid conditioner H06. Rinse it
off well.

INCI: Sodium Cocoyl Isethionate, Glycerin, Cocos Nucifera Oil,
Stearic Acid, Aloe Barbadensis Extract, Parfum, Potassium
Sorbate, Sodium Benzoate, Limonene, Linalool, Geraniol,
Citronellol, Citral.

EXTERNAL USE / KEEP IN A COOL DRY PLACE

Cosmética mediterránea con
aceite de almendras
Mediterranean cosmetics with almond oil

GENDERLESS / FOR ALL KIND OF HAIR

Esp /
Avanza y elige hoy para acondicionar tu cabello la opción más
efectiva y la más respetuosa con el medio ambiente por su tamaño reducido, cero plásticos y activos de origen natural que
garantizan un cabello hidratado, aumentando su sedosidad y
volumen.

¿Cómo utilizarlo? Frota el acondicionador con las manos y
aplica la ligera emulsión sobre el cabello, desde medios hasta
las puntas. Deja actuar unos minutos y enjuaga.

Eng /
Go ahead and choose today the most effective option for
conditioning your hair, and also the most respectful with our
environment because of its reduced size, zero plastics, and
natural origin actives that offer a hydrated hair increasing its
silkiness and volume.

¿How to use it? Gently rub the bar between your hands, and
apply the light cream onto your wet hair, distributing evenly on
the lengths. Leave it on for a few minutes and rinse.

INCI: Theobroma Cacao Seed Butter, Cetearyl Alcohol,
Behentrimonium Methosulfate, Prunus Amygdalus Dulcis Oil,
Butyrospermum Parkii Butter, Cetyl Alcohol, Butylene Glycol,
Parfum, Citral, Limonene, Geraniol.

EXTERNAL USE / KEEP IN A COOL DRY PLACE

Modesta
Cassinello

H06

Acondicionador Sólido
Solid Conditioner

Prunus dulcis

Dimensions:
135 mm × 65 mm × 65 mm

Material(s):
**Rives Shetland natural white
paper of Arjowiggins (350 gsm)**

Typeface(s):
Ogg

Print & Finishing:
Offset printing

● Pantone Black 7C

C: 63 Y: 64
M: 60 K: 65

○ C: 0
M: 0
Y: 0
K: 0

Jyugoya

Design Agency: **Frame inc.**
Art Direction: **Ryuta Ishikawa**
Design: **Ryuta Ishikawa, Kiyokazu Shimizu**
Client: **Jyugoya**

This project includes visual identity and packages of a confectionery store in front of a big *Shinto* shrine. Dedicated to the shrine, the design is based on *torii* (the gate of a Japanese shrine) and a symbol of rabbits.

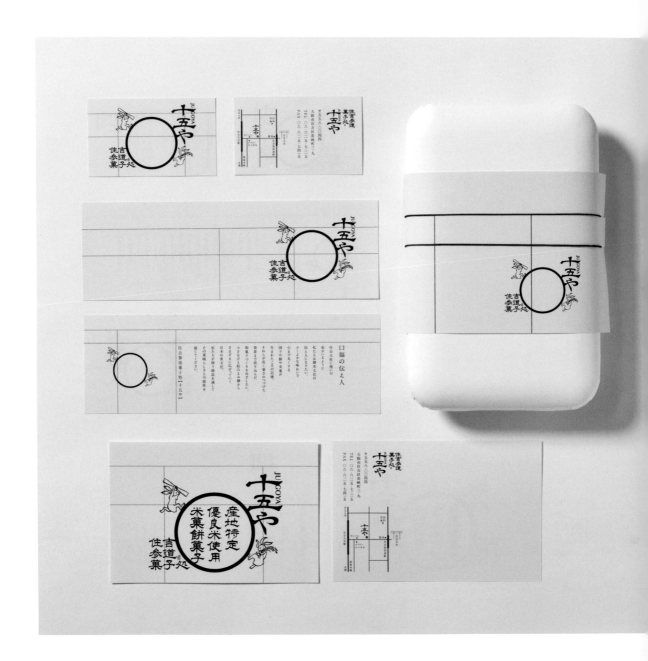

Dimensions:
75 mm × 75 mm × 150 mm
210 mm × 210 mm × 55 mm

Material(s):
Paper

○	●	●	●
C : 0	C : 24	C : 0	C : 0
M : 0	M : 18	M : 0	M : 100
Y : 0	Y : 40	Y : 0	Y : 100
K : 0	K : 0	K : 100	K : 0

Nadm Skincare

Design: **Renata Pereira**

The concept of the product is organic oil and extracts for the skin. The brand focuses on selected ingredients, quality and freshness. The design is developed around the universe of the Maghreb countries. The name, the typography and the colour palette are inspired by the landscapes of the Sahara—warm, soft and natural. For the graphic, Renata Pereira stayed in a minimalist style with clean lines and recycled paper.

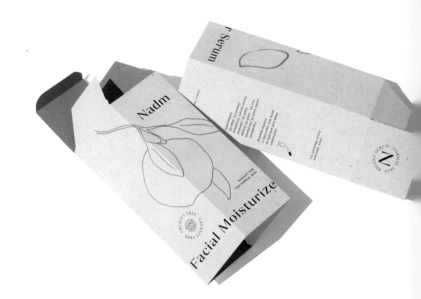

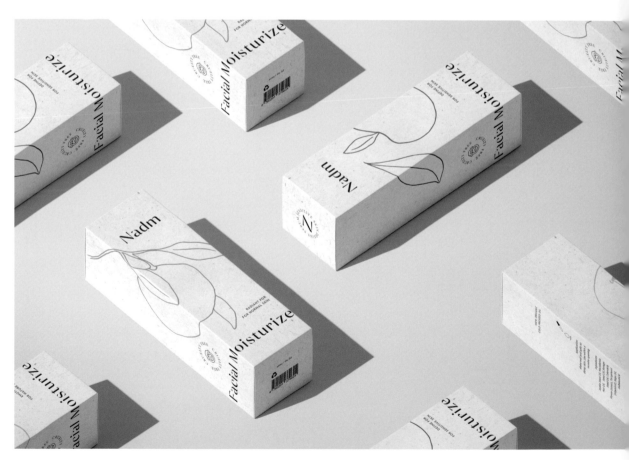

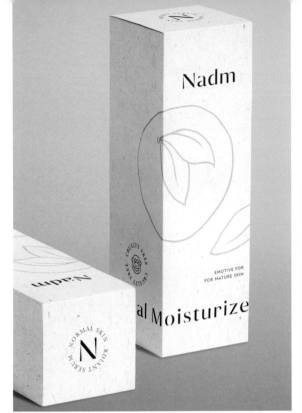

Dimensions:
40 mm × 40 mm × 125 mm

Material(s):
Recycled cotton paper

Typeface(s):
Gilda, Ponzu Regular, Cera Regular

Print & Finishing:
Matte uncoated finishing

C : 59
M : 57
Y : 0
K : 0

C : 6
M : 42
Y : 49
K : 0

C : 11
M : 19
Y : 18
K : 0

C : 8
M : 8
Y : 8
K : 0

SPRX Branding & Packaging

Design Agency: **AURG Design**
Client: **SPRX Korea**

SPRX is a brand of diet management and health supplement. As the brand's slogan is 'Tell a Story', AURG Design chose a simple pair of quotation marks as the logo that is used throughout the product line, expressing the brand's desire to have a conversation, to stress or indicate something important. Besides, the design team focused the renewed package design on a simple, clean yet urgent style. They highlighted the slogan as a signature quote and printed it in a special texture (hologram) to give it a more distinctive and unique look.

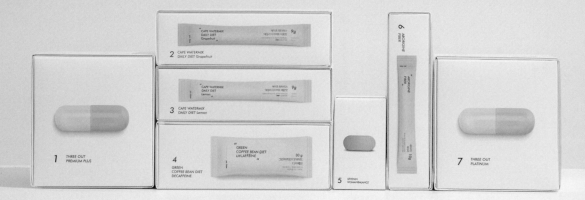

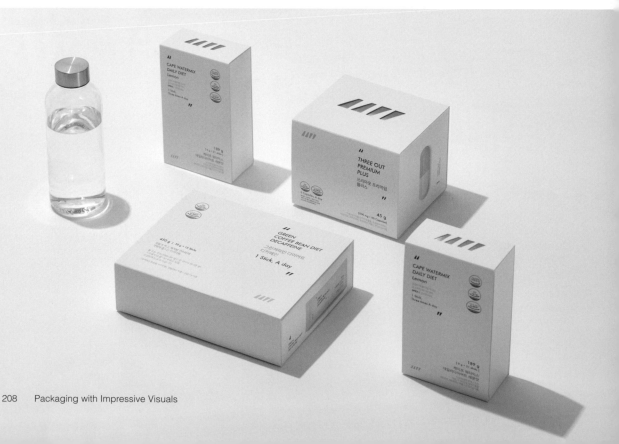

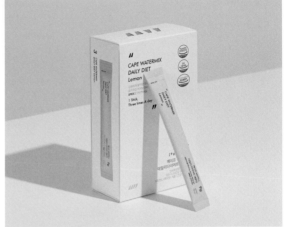

Dimensions:
A: 165 mm × 113 mm × 140 mm
B: 90 mm × 178 mm × 90 mm
C: 156 mm × 215 mm × 108 mm
D: 90 mm × 150 mm × 120 mm

Material(s):
**Paper, plastic container,
aluminium pouch**

Typeface(s):
Futura

Print & Finishing:
Hologram

Pantone 2915 C

C: 60   Y: 0
M: 9   K: 0

Pantone 7499 C

C: 1   Y: 24
M: 2   K: 0

Pantone 498 C

C: 0   Y: 21
M: 20   K: 0

Pantone 621 C

C: 12   Y: 12
M: 1   K: 2

TE&R Packaging Design

TE&R is a care product for strengthening the skin's ability to repair itself and promote regrowth. The designer drew inspiration from the shape of lactoferrin which is one of the important ingredients in this product.

Design: **Echo Yang**
Photography: **Echo Yang**
Printing: **Wei-Yang Print Plan Agent**

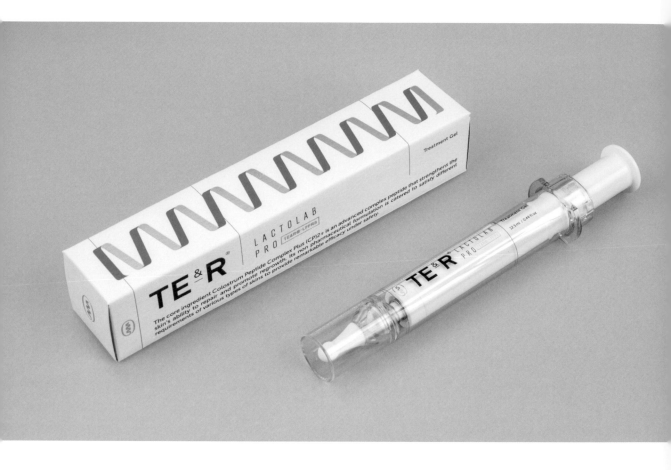

Dimensions:
33 mm × 33 mm × 170 mm

Material(s):
Paper

Typeface(s):
Museo Sans

Print & Finishing:
Offset printing, hot-stamping

Black
C : 0   Y : 0
M : 0   K : 100

Pantone Orange 021C
C : 0   Y : 100
M : 74   K : 0

Highline Wellness

Design Agency: **Unspoken Agreement**
Design: **Saxon Cambell, Spencer Woolcott**

Highline Wellness provides natural, affordable and effective Cannabidiol supplements. The company approached Unspoken Agreement at a critical juncture poised for rapid growth and boasting a fanbase of loyal users but not packaged for the big leagues. The design team's goal was to elevate the company to the next level, while still honouring its customers. They went about redesigning the brand from the ground up—a new, more sophisticated identity, fresh and welcoming brand voice, airier colour palette, clean packaging and a more modern, friendly website.

Dimensions:
108 mm × 60 mm × 60 mm

Material(s):
Paper (cotton texture)

Typeface(s):
Clearface, Sharp Sans

Print & Finishing:
Four-colour printing, debossing, hot-stamping

C: 78
M: 28
Y: 24
K: 1

C: 99
M: 71
Y: 43
K: 32

Michlberger Small Batch Fruit Spirit

Design: **Zoltán Visnyai**
Photography: **Márton Ács**

The colours match with the flavours of the fruits, the gradient reflects the process of distillation, and the lacquered pattern refers to traditional etching. There are two rubber bands attaching the label so that it is removable and functions as a decoration with a short description of the traditions and culture of pálinka consumption and the steps of small batch distillation.

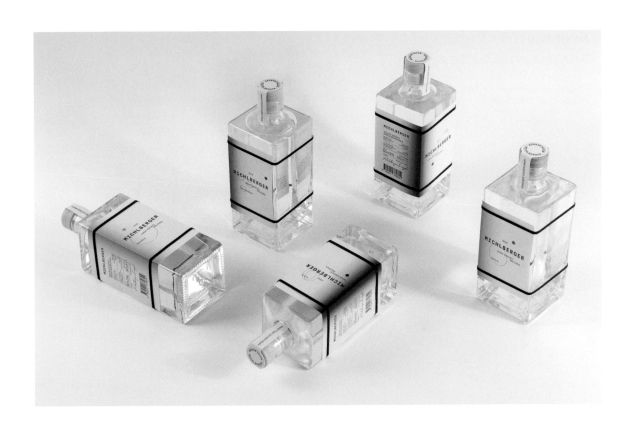

Dimensions:
75 mm × 75 mm × 200 mm

Material(s):
**Matte carton, paper,
rubber band, glass**

Typeface(s):
Suisse Int'l Mono, Suisse Neue

Print & Finishing:
Digital printing, lacquering

C: 0   C: 17
M: 8   M: 100
Y: 4   Y: 100
K: 0   K: 9

C: 0   C: 17
M: 0   M: 51
Y: 7   Y: 100
K: 10   K: 5

C: 15   C: 100
M: 6   M: 53
Y: 0   Y: 28
K: 0   K: 14

C: 9   C: 100
M: 10   M: 100
Y: 0   Y: 0
K: 0   K: 0

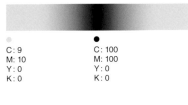

C: 0   C: 0
M: 8   M: 70
Y: 10   Y: 89
K: 0   K: 0

Scence: The Future of Sustainable Skincare

Design Agency: **Kingdom & Sparrow**
Photography: **Howard Oates**
Client: **Scence Skincare**

Scence commissioned Kingdom & Sparrow to grow their all-natural, completely sustainable skincare brand and update their brand for modern, discerning and eco-conscious consumers. Therefore, the design agency developed a fresh look to bring the brand into the modern, commercial skincare market. They created a strong identity with more ownable flourishes and contemporary design. Also, they chose a mature colour palette and hand-painted illustrations to reflect the natural ingredients and add a premium touch. Clear visual navigation and range architecture to help consumers differentiate between products and scents were also paramount.

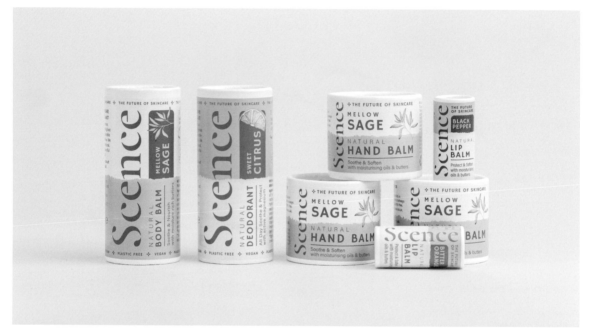

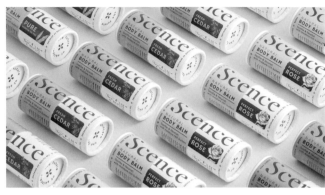

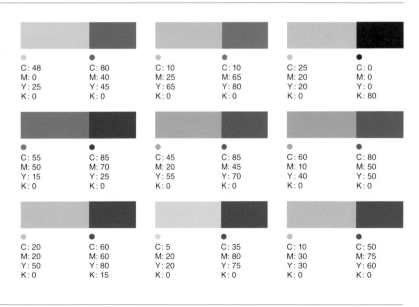

Dimensions:
Body balm: 96 mm × 40 mm (diametre)
Hand cream: 46 mm × 52 mm (diametre)
Face balm: 46 mm × 46 mm (diametre)
Lip balm: 48 mm × 22 mm (diametre)

Material(s):
Paper tubes and pots

Typeface(s):
Hammersmith One,
Cormorant Garamond,
Europa Regular

Print & Finishing:
Digital printing using
vegan inks

C: 48 / M: 0 / Y: 25 / K: 0	C: 80 / M: 40 / Y: 45 / K: 0	C: 10 / M: 25 / Y: 65 / K: 0	C: 10 / M: 65 / Y: 80 / K: 0	C: 25 / M: 20 / Y: 20 / K: 0	C: 0 / M: 0 / Y: 0 / K: 80
C: 55 / M: 50 / Y: 15 / K: 0	C: 85 / M: 70 / Y: 25 / K: 0	C: 45 / M: 20 / Y: 55 / K: 0	C: 85 / M: 45 / Y: 70 / K: 0	C: 60 / M: 10 / Y: 40 / K: 0	C: 80 / M: 50 / Y: 50 / K: 0
C: 20 / M: 20 / Y: 50 / K: 0	C: 60 / M: 60 / Y: 80 / K: 15	C: 5 / M: 20 / Y: 20 / K: 0	C: 35 / M: 80 / Y: 75 / K: 0	C: 10 / M: 30 / Y: 30 / K: 0	C: 50 / M: 75 / Y: 60 / K: 0

3rd Universe

Design Agency: **Transwhite Studio**
Art Direction: Yu Qiongjie
Design: **Yang Fan, Zhao Tingyu**
Type Design: **Yang Fan, Zhao Tingyu**
Photography: **Fu Ruilin, 3rd Universe**
Client: **3rd Universe**

3rd Universe is a new revolutionary cosmetics brand that creates advanced makeup for a youthful, independent and innovative audience. Transwhite Studio captured the motions of pulling repeated every time the makeup is applied. The graphic language of pulling, represented by a stretching letter *d*, is employed throughout the identity with the idea of opening a new world to the customer. Different graphics are created and embedded in the letter *d* to match each product. To extend the concept of pulling, functional sub-logos were designed and applied to specific packaging. The three spot colours used for the eyeshadow packaging correspond to the colours of the products themselves so that consumers can directly distinguish the product from the packaging.

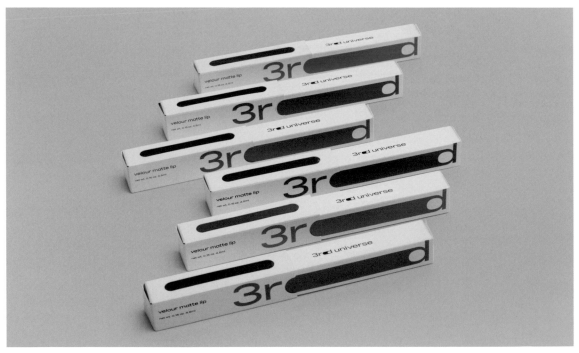

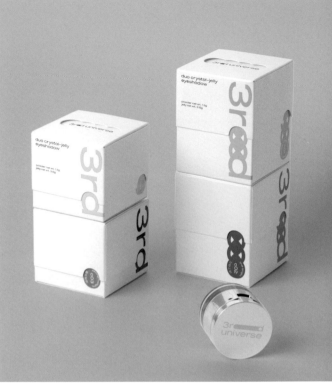

Dimensions:
Lipstick package:
20 mm × 20 mm × 95 mm
Eyeshadow package:
56 mm × 55 mm × 55 mm

Material(s):
Paper

Typeface(s):
Custom typeface

Print & Finishing:
Die-cutting, relief printing,
UV sealing

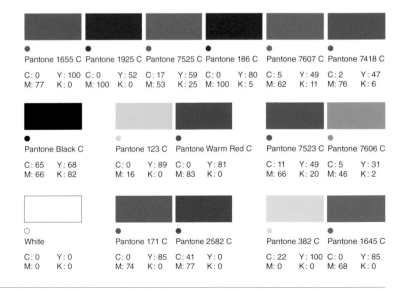

Pantone 1655 C	Pantone 1925 C	Pantone 7525 C	Pantone 186 C	Pantone 7607 C	Pantone 7418 C
C: 0 Y: 100	C: 0 Y: 52	C: 17 Y: 59	C: 0 Y: 80	C: 5 Y: 49	C: 2 Y: 47
M: 77 K: 0	M: 100 K: 0	M: 53 K: 25	M: 100 K: 5	M: 62 K: 11	M: 76 K: 6

Pantone Black C	Pantone 123 C	Pantone Warm Red C	Pantone 7523 C	Pantone 7606 C
C: 65 Y: 68	C: 0 Y: 89	C: 0 Y: 81	C: 11 Y: 49	C: 5 Y: 31
M: 66 K: 82	M: 16 K: 0	M: 83 K: 0	M: 66 K: 20	M: 46 K: 2

White	Pantone 171 C	Pantone 2582 C	Pantone 382 C	Pantone 1645 C
C: 0 Y: 0	C: 0 Y: 85	C: 41 Y: 0	C: 22 Y: 100	C: 0 Y: 85
M: 0 K: 0	M: 74 K: 0	M: 77 K: 0	M: 0 K: 0	M: 68 K: 0

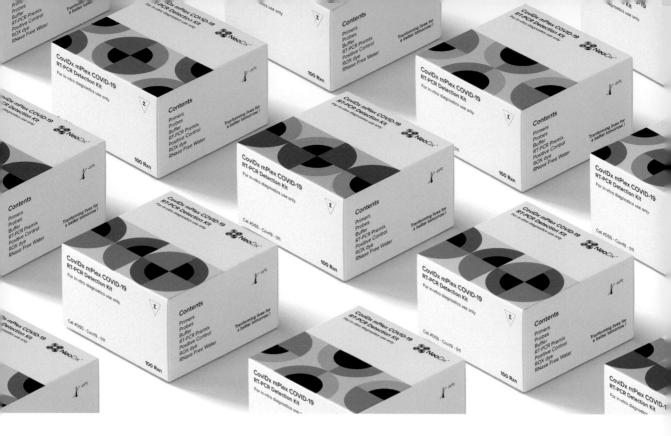

NeoDx Biotech

Design Agency: **Mechi Co. Design**
Art Direction: **Meroo Seth**
Creative Direction: **Nachiket Jadhav**
Design: **Meroo Seth**

NeoDx is a brand of molecular diagnostics, specialising in testing tools that help in giving accurate results about several health issues. Mechi's challenge was to design the look of the new brand that is less intimidating within the serious field without losing its quality. The whole concept came from a circle as the symbol of the cycle of life and transformation. Based on that, they drew parallels with the DNA helix and crystallised structure to form the logo with a colour palette inspired by the four main components of DNA. The combination of white packaging and Neo's vibrant colours forms a beautiful, clean and precise brand that personifies Neo's expertise in creating their testing kits.

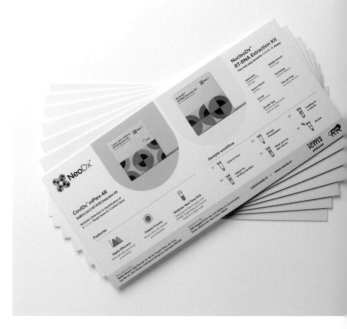

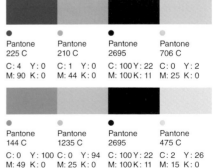

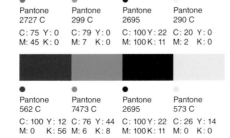

Dimensions:
115 mm × 98 mm × 55 mm

Material(s):
Paperboard

Typeface(s):
**Proxima Nova,
Gotham Rounded**

Print & Finishing:
**Digital printing,
matte finishing**

Pantone
2727 C
C: 75 Y: 0
M: 45 K: 0

Pantone
299 C
C: 79 Y: 0
M: 7 K: 0

Pantone
2695
C: 100 Y: 22
M: 100 K: 11

Pantone
290 C
C: 20 Y: 0
M: 2 K: 0

Pantone
225 C
C: 4 Y: 0
M: 90 K: 0

Pantone
210 C
C: 1 Y: 0
M: 44 K: 0

Pantone
2695
C: 100 Y: 22
M: 100 K: 11

Pantone
706 C
C: 0 Y: 2
M: 25 K: 0

Pantone
562 C
C: 100 Y: 12
M: 0 K: 56

Pantone
7473 C
C: 76 Y: 44
M: 6 K: 8

Pantone
2695
C: 100 Y: 22
M: 100 K: 11

Pantone
573 C
C: 26 Y: 14
M: 0 K: 0

Pantone
144 C
C: 0 Y: 100
M: 49 K: 0

Pantone
1235 C
C: 0 Y: 94
M: 25 K: 0

Pantone
2695
C: 100 Y: 22
M: 100 K: 11

Pantone
475 C
C: 2 Y: 26
M: 15 K: 0

The Singular Olivia - Solid Shampoo

Design Agency: **MARINA GOÑI STUDIO**
Design: **Marina Goñi, Ane Garmendia, Isabel Pérez**

The Singular Olivia is a cosmetic brand that creates and develops products and also markets special products of other brands. For this collection of five solid shampoos, MARINA GOÑI STUDIO came up with an idea to design the boxes—each shampoo's ingredients should evoke an imaginary country and landscape. Thus, they built a continuous landscape going through those five countries. The names of the shampoos in gold stamping on the front side of each box provide unmistakable elegance to the brand. The final result is a collection of five solid, high-quality shampoos in cheerful and colourful packaging with a sophisticated printed cardboard structure.

Dimensions:
680 mm × 680 mm × 770 mm

Material(s):
Cardboard

Print & Finishing:
Digital foiling

C: 11	C: 34	C: 56	C: 29
M: 19	M: 2	M: 27	M: 12
Y: 54	Y: 19	Y: 40	Y: 42
K: 1	K: 0	K: 9	K: 0

C: 0	C: 4	C: 7	C: 15
M: 48	M: 26	M: 38	M: 42
Y: 28	Y: 15	Y: 40	Y: 56
K: 0	K: 0	K: 0	K: 4

C: 44	C: 31	C: 52	C: 19
M: 3	M: 14	M: 31	M: 28
Y: 20	Y: 32	Y: 51	Y: 33
K: 0	K: 1	K: 7	K: 4

C: 44	C: 11	C: 15	C: 7
M: 14	M: 19	M: 49	M: 36
Y: 47	Y: 0	Y: 65	Y: 56
K: 1	K: 0	K: 3	K: 0

C: 16	C: 11	C: 30	C: 53
M: 42	M: 19	M: 42	M: 28
Y: 47	Y: 16	Y: 46	Y: 40
K: 4	K: 0	K: 20	K: 10

C: 34	C: 4	C: 31	C: 11	C: 11
M: 2	M: 26	M: 14	M: 19	M: 19
Y: 19	Y: 15	Y: 32	Y: 0	Y: 16
K: 0	K: 0	K: 1	K: 0	K: 0

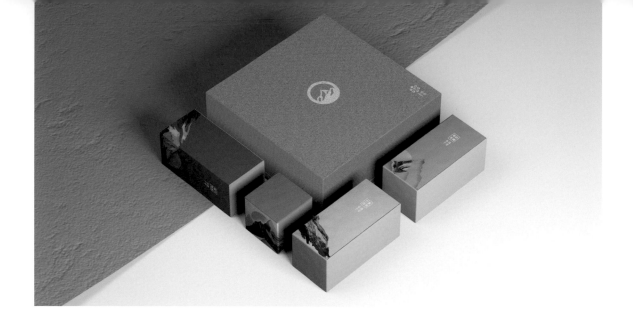

Wu Di Cha
Cold Brew Tea

Design: **Wing Yang**
Copywriting: **Longlong**

The graphics of mountains on the package pay homage to the traditional and time-honoured drinking of tea and, at the same time, the diverse colours bring a fresh touch to illustrate the overthrowing of the traditional way of brewing tea with hot water.

Dimensions:
90 mm × 198 mm

Material(s):
Specialty paper

Print & Finishing:
Four-colour printing

C: 90	C: 22
M: 75	M: 76
Y: 37	Y: 40
K: 2	K: 0

C: 53	C: 78	C: 80
M: 21	M: 56	M: 75
Y: 71	Y: 73	Y: 72
K: 0	K: 17	K: 48

C: 7	C: 14	C: 80
M: 9	M: 43	M: 75
Y: 22	Y: 74	Y: 72
K: 0	K: 0	K: 48

C: 22	C: 7	C: 80
M: 76	M: 9	M: 75
Y: 40	Y: 22	Y: 72
K: 0	K: 0	K: 48

C: 0	C: 7	C: 80	C: 48
M: 55	M: 9	M: 75	M: 87
Y: 65	Y: 22	Y: 72	Y: 100
K: 0	K: 0	K: 48	K: 20

C: 90	C: 22	C: 14	C: 53
M: 75	M: 76	M: 43	M: 21
Y: 37	Y: 40	Y: 74	Y: 71
K: 2	K: 0	K: 0	K: 0

Moss Packaging - Limited Edition on Exhibition

Design: **Bacon Liu**
Photography: **Bacon Liu**

Moss is a terrarium plant brand of custom potted plants. As mosses reproduce themselves through spores explosion, the designer tried to deliver the concept of breeding through a colourfully 'sexy' packaging design with a palette of neon colours and references to Chinese neon signboard at night. He chose 70% recycled pulp paper and applied silkscreen printing, hot-stamping and debossing to finish the bright and high-contrast packaging. In addition, the typography illustrates mushroom clouds to extend the concept of spores explosion.

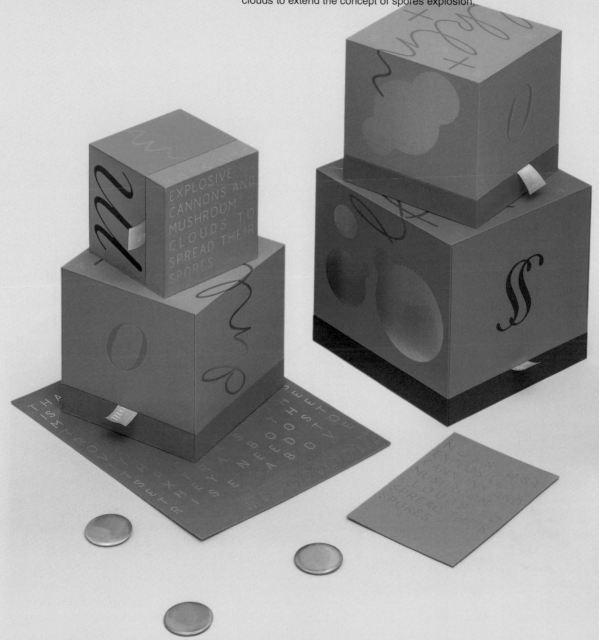

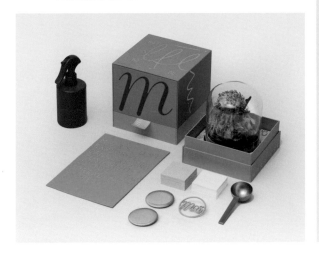

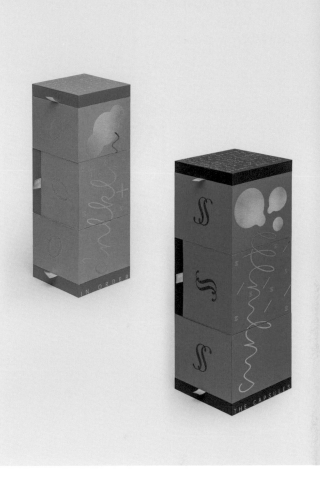

Dimensions:
A: 196 mm × 196 mm × 196 mm
B: 251 mm × 251 mm × 251 mm
C: 356 mm × 356 mm × 356 mm
D: 386 mm × 386 mm × 386 mm

Material(s):
Paperboard rigid box,
paper of TAKEO_NT RASHA

Typeface(s):
Times Italic

Print & Finishing:
Silkscreen printing,
debossing, hot-stamping
with gold foil

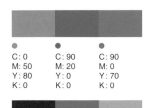

C : 90	C : 0	C : 100
M : 0	M : 50	M : 70
Y : 70	Y : 80	Y : 0
K : 0	K : 0	K : 0

C : 0	C : 90	C : 90
M : 50	M : 20	M : 0
Y : 80	Y : 0	Y : 70
K : 0	K : 0	K : 0

C : 90	C : 50	C : 20
M : 20	M : 90	M : 100
Y : 0	Y : 0	Y : 20
K : 0	K : 0	K : 0

C : 50	C : 90	C : 10
M : 90	M : 10	M : 30
Y : 0	Y : 40	Y : 100
K : 0	K : 0	K : 0

W-Moon-Y
Muti-Vitabin

Art Direction: **Hunk Xing**
Creative Direction: **Raven Lee**
Direction: **Cash Wu**
Design: **Hunk Xing**
Copywriting: **Binbin**
Client: **WMY WORKS**

This set of mooncakes includes nine flavours corresponding to nine types of 'vitamins' for young people to cope with the nine typical predicaments in their daily life such as staying up late, over-drinking, homesickness and so on. In this packaging design, the design team converted their good wishes into the concept of vitamin in a humourous way, extracted the visual elements from pharmaceutical products and chose a bright and vibrant colour palette to cheer the consumers up.

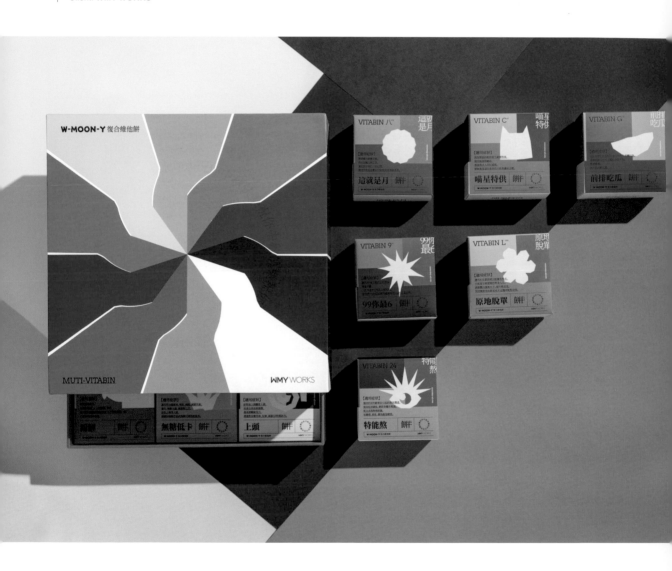

Dimensions:
Outer box:
268 mm × 268 mm × 70 mm
Small box:
85 mm × 85 mm × 50 mm

Material(s):
Cardboard, coated paper,
matte art paper

Typeface(s):
Source Han Serif

Print & Finishing:
Four-colour printing,
UV ink

C: 0	C: 29	C: 94	C: 100	C: 75	C: 0	C: 5	C: 19	C: 0
M: 88	M: 38	M: 74	M: 95	M: 78	M: 64	M: 36	M: 0	M: 77
Y: 73	Y: 99	Y: 2	Y: 26	Y: 0	Y: 10	Y: 86	Y: 79	Y: 89
K: 0	K: 0	K: 0	K: 0	K: 0	K: 0	K: 0	K: 0	K: 0

C: 98	C: 67	C: 0	C: 2	C: 51	C: 0	C: 23	C: 2	C: 75
M: 82	M: 0	M: 77	M: 87	M: 0	M: 64	M: 60	M: 53	M: 78
Y: 0	Y: 29	Y: 89	Y: 23	Y: 36	Y: 10	Y: 0	Y: 81	Y: 0
K: 0	K: 0	K: 0	K: 0	K: 0	K: 0	K: 0	K: 0	K: 0

C: 0	C: 67	C: 100	C: 55	C: 74	C: 5	C: 81	C: 32	C: 19
M: 93	M: 44	M: 95	M: 10	M: 25	M: 36	M: 79	M: 97	M: 0
Y: 55	Y: 0	Y: 26	Y: 12	Y: 71	Y: 86	Y: 83	Y: 15	Y: 79
K: 0	K: 0	K: 0	K: 0	K: 0	K: 0	K: 66	K: 0	K: 0

C: 8	C: 75	C: 0	C: 80	C: 71	C: 29	C: 62	C: 0	C: 94
M: 1	M: 78	M: 88	M: 68	M: 100	M: 38	M: 0	M: 82	M: 74
Y: 82	Y: 0	Y: 73	Y: 10	Y: 49	Y: 99	Y: 100	Y: 42	Y: 2
K: 0	K: 0	K: 0	K: 0	K: 12	K: 0	K: 0	K: 0	K: 0

The ROCCA
Biscuit Collection

Design Agency: **WWAVE DESIGN**
Design: **Amy Un Cho Ian, Ken Ho Ion Fat, Kenneth Ho**
Photography: **Andrew Kan**

The ROCCA Biscuit Collection features six boxes of biscuits of different flavours, shapes and texture. WWAVE kept the biscuits compact in a delicate metal box wrapped by coloured paper sleeves with the key visual hot-stamped on, giving it a classy touch. The key visual is type-oriented, matching each flavour with a unique typeface paired with a specific colour and paper texture. For example, Gill Sans is used on peach pink smooth paper for 'Rose sablée tea biscuits' reminiscent of an elegant English afternoon tea. They wished to convey a direct message to the customers by having the flavours as the design elements and to explore the customers' interpretations towards the design.

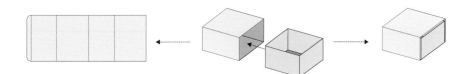

Dimensions:
100 mm × 70 mm × 80 mm

Material(s):
Metal, six types of paper

Print & Finishing:
Debossing, embossing, hot-stamping with gold foil

C: 30	C: 0	C: 88	C: 50	C: 9	C: 18
M: 0	M: 70	M: 76	M: 56	M: 20	M: 63
Y: 26	Y: 52	Y: 40	Y: 55	Y: 23	Y: 70
K: 0	K: 0	K: 35	K: 74	K: 3	K: 7

BANBOU MOONCAKE

Design Agency: **Y.STUDIO**
Art Direction: **Ziji Yu**
Design Direction: **Jun Guo**
Design: **Nan Lin, Kangtian Quan**
Client: **BANBOU PATISSERIE & COFFEE**

Instead of the moon, Y.STUDIO chose
stars as the main visuals. Different from
the traditional mooncake packages,
the designers employed a bright colour
palette of white and different shades
of blue, giving the product a sense of
freshness.

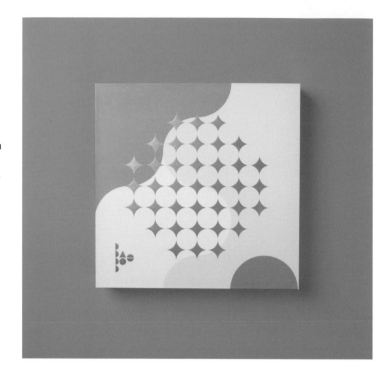

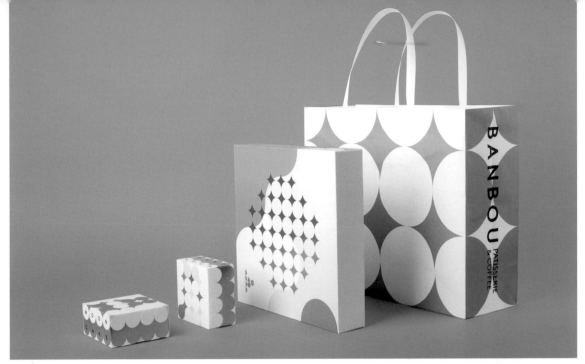

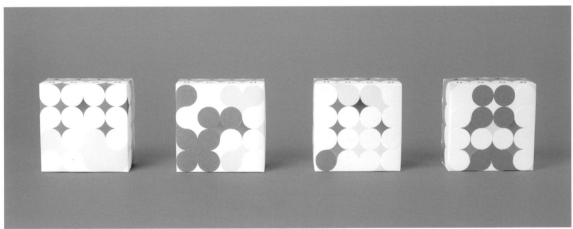

Dimensions:
225 mm × 225 mm × 54 mm

Material(s):
Paper

Typeface(s):
Brown, Source Han Serif

Print & Finishing:
Spot-colour printing, hot-stamping with matte gold foil, embossing

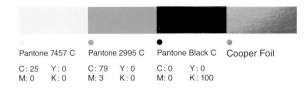

Pantone 7457 C
C: 25   Y: 0
M: 0   K: 0

Pantone 2995 C
C: 79   Y: 0
M: 3   K: 0

Pantone Black C
C: 0   Y: 0
M: 0   K: 100

Cooper Foil

DAS kafeD

Design Agency: **Victor Branding Design**
Client: **Ju Ho Feng Co., Ltd.**

As usual, DAS kafeD chooses desserts with the strictest standard. Dessert is just like an artwork that is satisfying not only in terms of taste but also visuals. The outer box is presented in blue with bronze foil hot-stamped, giving a sense of low-key elegance. In addition, the special texture on the paper specially chosen makes the packaging meticulously warm. For the inner box, the main visual element is brush strokes.

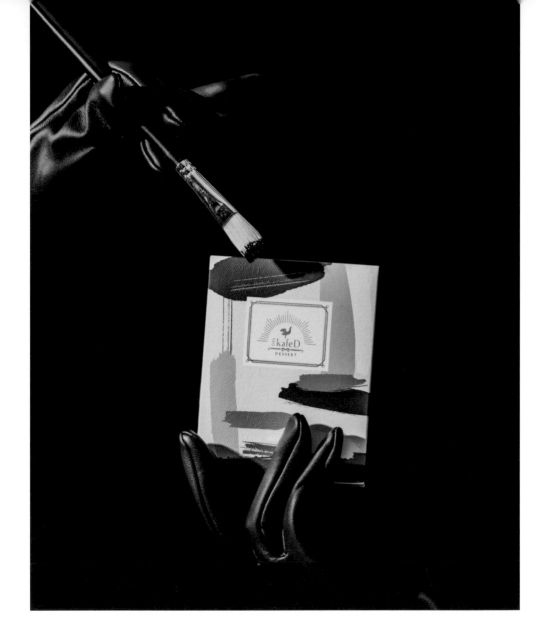

Dimensions:
Outer box:
305 mm × 320 mm × 65 mm
Inner box:
120 mm × 100 mm × 62 mm

Material(s):
MetsäBoard Natural FBB with matte coating

Print & Finishing:
Hot-stamping with bronze foil

Pantone 7684 C

C: 85   Y: 18
M: 65   K: 0

C: 55
M: 40
Y: 15
K: 0

C: 0
M: 35
Y: 35
K: 0

C: 5
M: 5
Y: 5
K: 0

Bakes Mooncake

Design Agency: **The Lab Saigon**
Creative Direction: **Tuan Le**
Graphic Design: **Tran Nguyen, Trang Dinh**
Illustration: **Hung Le**
3D: **Naomi Nguyen**
Project Management: **Phuong Anh Nguyen, Huyen Vo, Khoa Do**
Production Management: **Hanh Le**
Photography: **Duong Gia Hieu**

Bakes is a pastry brand in Vietnam, and The Lab Saigon designed the mooncake packaging for it. They took inspiration from the moon's reflection on a lake, a reference to Bakes' flagship store in the historic Turtle Lake neighborhood. The ripple pattern adorns the top of the cylindrical box and the mooncake itself (created from a custom 3D-printed cake mold). Bespoke illustrations reimagine the story of each flavour, wrapping around the cylinder (you can spin the box and see a stop motion retelling of the story). Finally, graphic patterns are taken directly from the historical architecture of the Turtle Lake.

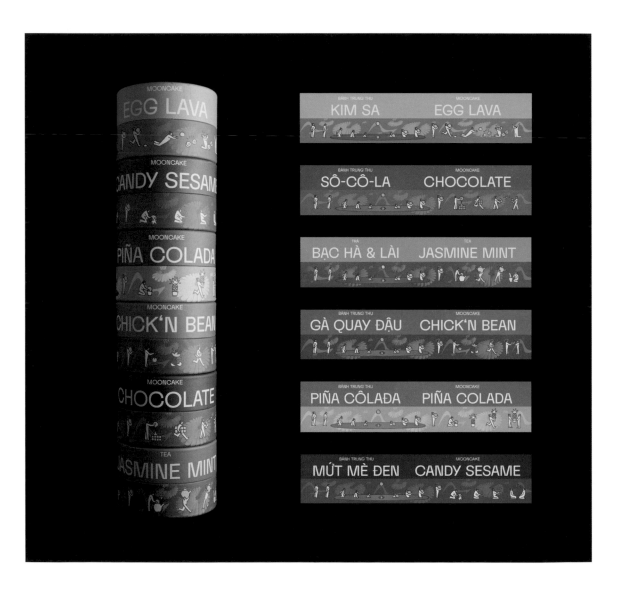

Dimensions:
60 mm × 85 mm (diametre)

Material(s):
**Constellation snow E/E49
country (130 gsm)**

Print & Finishing:
Offset printing

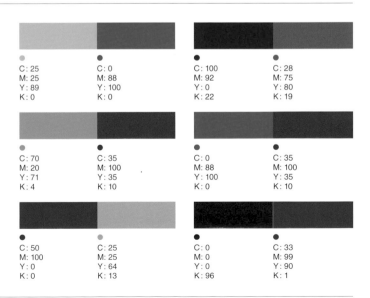

C: 25
M: 25
Y: 89
K: 0

C: 0
M: 88
Y: 100
K: 0

C: 100
M: 92
Y: 0
K: 22

C: 28
M: 75
Y: 80
K: 19

C: 70
M: 20
Y: 71
K: 4

C: 35
M: 100
Y: 35
K: 10

C: 0
M: 88
Y: 100
K: 0

C: 35
M: 100
Y: 35
K: 10

C: 50
M: 100
Y: 0
K: 0

C: 25
M: 25
Y: 64
K: 13

C: 0
M: 0
Y: 0
K: 96

C: 33
M: 99
Y: 90
K: 1

YAMAMOTOYAMA

Design Agency: **NOSIGNER**
Art Direction: **Eisuke Tachikawa**
Design: **Eisuke Tachikawa, Ryota Mizusako, Nozomi Aoyama**
Photography: **Kunihiko Sato**
Client: **YAMAMOTOYAMO Co., Ltd.**

NOSIGNER worked on rebranding a project called 'YAMAMOTOYAMA', a long-established tea merchant that has supported Japanese culture and the aesthetic sense of the Edo period. With the concept of 'Return to the origin of Edo', the YAMAMOTOYAMA original small crests and the calligraphy style of Edo were adopted. The designers also redesigned the packaging to be more modern and elegant while retaining the charm of the long history brand by referring to the traditional colours and structure of the scrolls during the Edo period. This process has led to a gradual and successful shift to a new design.

Dimensions:
ASAKUSA box
(seaweed set of two):
87 mm × 240 mm × 185 mm
Tea box:
120 mm × 78 mm × 68 mm

Material(s):
Paper, iron

Typeface(s):
AXIS Condensed Std M,
A-OTF A1Mincho Std Bold,
MS gothic, Quincy CF Text,
A-OTF Shuei Mincho Pro B

Print & Finishing:
Four-colour printing,
spot-colour printing

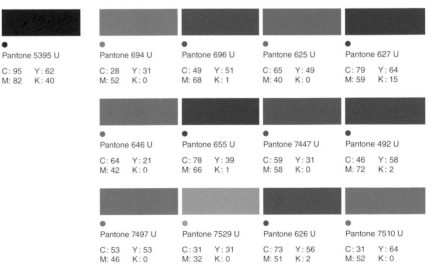

Pantone 5395 U	Pantone 694 U	Pantone 696 U	Pantone 625 U	Pantone 627 U
C : 95 Y : 62 M : 82 K : 40	C : 28 Y : 31 M : 52 K : 0	C : 49 Y : 51 M : 68 K : 1	C : 65 Y : 49 M : 40 K : 0	C : 79 Y : 64 M : 59 K : 15

Pantone 646 U	Pantone 655 U	Pantone 7447 U	Pantone 492 U
C : 64 Y : 21 M : 42 K : 0	C : 78 Y : 39 M : 66 K : 1	C : 59 Y : 31 M : 58 K : 0	C : 46 Y : 58 M. 72 K : 2

Pantone 7497 U	Pantone 7529 U	Pantone 626 U	Pantone 7510 U
C : 53 Y : 53 M : 46 K : 0	C : 31 Y : 31 M : 32 K : 0	C : 73 Y : 56 M : 51 K : 2	C : 31 Y : 64 M : 52 K : 0

Chu Dessert: Brand Design with Poetic Flavor

Design Agency: **StudioPros**
Art Direction: **Yi-Hsuan Li**
Design: **Yi-Hsuan Li, Jane Lu**
Naming and Concept: **Kai-Hsiang Ho**
Photography: **Shengyuan Hsu**
Client: **Chu Dessert**

The brand Chu Dessert was named after its founder and was developed and motivated by the poem 'Ode to the Chrysanthemum' written by a renowned poet in ancient China—Bai Juyi. The brand logo was designed by a new sculptor Chen Zu-Bei who drew the Chinese character ' 菊 ' ('Chu', chrysanthemum) with smooth curves to portray the image of the flower. The abstract patterns of the scattering petals are decorated with white and gold, echoing the poem that mentions frost on the flowers and golden buds. The package is presented in three colours creating different ambiances for different scenarios—the green and white ones are usually for gifts while the red one for weddings.

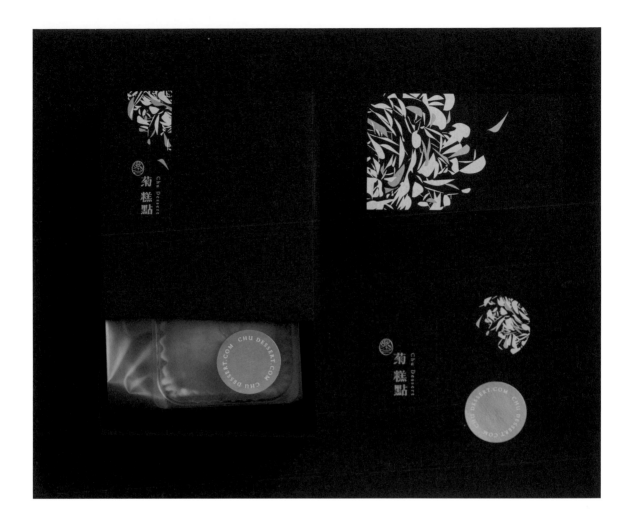

Dimensions:
100 mm × 100 mm × 100 mm

Material(s):
Paper

Typeface(s):
Times New Roman, A-OTF A1 Mincho

Print & Finishing:
Hot-stamping with gold foil

● C: 75 M: 68 Y: 91 K: 48	● C: 40 M: 50 Y: 100 K: 0	○ C: 0 M: 0 Y: 0 K: 0

○ C: 10 M: 8 Y: 12 K: 0	● C: 40 M: 50 Y: 100 K: 0	○ C: 0 M: 0 Y: 0 K: 0

● C: 30 M: 94 Y: 84 K: 33	● C: 40 M: 50 Y: 100 K: 0	○ C: 0 M: 0 Y: 0 K: 0

● C: 75 M: 68 Y: 91 K: 48	○ C: 10 M: 8 Y: 12 K: 0	● C: 30 M: 94 Y: 84 K: 33

SOLAR SYSTEM PACKAGING

Design Agency: **K9 Design**
Creative Direction: **Kevin Wei-Cheng Lin**
Design: **Kevin Wei-Cheng Lin, Robert Yang**
Photography: **Férguson Chang**

K9 Design's approach to this packaging of a mooncake brand was combining a western designing concept with this traditional Chinese dessert to attract young people. The packaging illustrates the solar system to convey the importance to cherish gifts from the remote universe (like the moon). Gold and platinum are the main colours of the packaging, which the design team believed would present the consumers a humourous image of contemporary art. As they positioned the packaging between premium products and collections, they chose metallic paper and applied hot-stamping to specific areas of the boxes to add a layer of details to it. To differentiate the products, different types of colour paper were used for the consumers to identify them quickly.

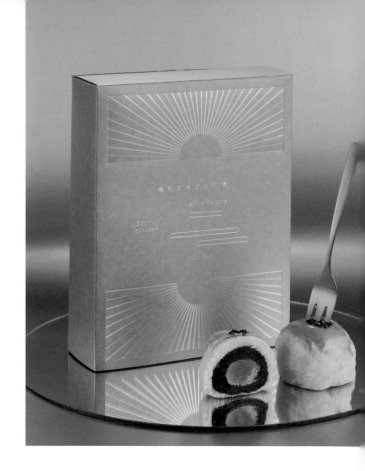

Dimensions:
210 mm × 160 mm × 53 mm

Material(s):
Paper from Alchemy series of Arjowiggins

Print & Finishing:
Hot-stamping with silver and gold foils

Pantone 8640 C

C: 47   Y: 95
M: 53   K: 2

Pantone 877 C

C: 45   Y: 34
M: 34   K: 0

Core & Rind:
Brand Identity
and Packaging Design

Design Agency: **Herefor Studio**
Creative Direction: **Cory Uehara, Ryan Hammond**
Management: **Jena Garlinghouse**

Core & Rind (Candi & Rita) are two
cheesy goddesses on a mission to
provide the world with plant-based,
cheesy goodness and rid the centre aisle
of bad imitation cheese. Herefor Studio
developed a bold new logo, gooey waves
of goodness that shift from flavour to
flavour, and an icon system that helps
drive product use education. The result
brings to life the game-changing equities
of the product—it's cheese, but made
from plant!

Dimensions:
76 mm × 83 mm (diametre)

Material(s):
Glass jar

Typeface(s):
Tropiline, GT America

○ 20% Pantone 155C	● Pantone Black 6 C	○ Pantone 137 C	Pantone Yellow C	○ Pantone 636 C	● Pantone 1505 C
C: 9 Y: 40 M: 23 K: 0	C: 100 Y: 32 M: 61 K: 96	C: 0 Y: 100 M: 36 K: 0	C: 0 Y: 100 M: 1 K: 0	C: 42 Y: 0 M: 0 K: 0	C: 0 Y: 100 M: 64 K: 0

○ 20% Pantone 155C	● Pantone Black 6 C	○ Pantone 137 C	Pantone Yellow C	○ Pantone 176 C	● Pantone 7622 C
C: 9 Y: 40 M: 23 K: 0	C: 100 Y: 32 M: 61 K: 96	C: 0 Y: 100 M: 36 K: 0	C: 0 Y: 100 M: 1 K: 0	C: 0 Y: 9 M: 43 K: 0	C: 0 Y: 77 M: 98 K: 37

○ 20% Pantone 155C	● Pantone Black 6 C	○ Pantone 137 C	● Pantone 4975 C	● Pantone 172 C	○ Pantone 176 C
C: 9 Y: 40 M: 23 K: 0	C: 100 Y: 32 M: 61 K: 96	C: 0 Y: 100 M: 36 K: 0	C: 27 Y: 62 M: 90 K: 83	C: 0 Y: 98 M: 80 K: 0	C: 0 Y: 9 M: 43 K: 0

○ 20% Pantone 155C	● Pantone Black 6 C	○ Pantone 9221 C	Pantone 3248 C	○ Pantone 636 C	● Pantone 2767 C
C: 9 Y: 40 M: 23 K: 0	C: 100 Y: 32 M: 61 K: 96		C: 52 Y: 32 M: 0 K: 0	C: 42 Y: 0 M: 0 K: 0	C: 100 Y: 0 M: 71 K: 66

PINEAPPLE COLLABORATIVE
Olive Oil

Design Agency: **STUDIO L'AMI**
Photography: **STUDIO L'AMI**
Client: **Pineapple Collaborative**

Pineapple Collaborative is on a mission to connect and celebrate women who love food. For the olive oil, STUDIO L'AMI tapped into the nostalgic feeling associated with the classic olive oil tin which is presented in three contemporary colours for the consumers to choose the one best complementing their kitchens. This ultimately extends from a brand philosophy that says 'what we cook and eat reflects our unique styles, identities and values.' The family of olive oils is presented alongside the apple-cider vinegar in an apothecary glass bottle with a natural feeling wood or cork top. The design team also introduced new and distinctive typefaces to establish a family look and feel to express the product's naming and messaging through typography.

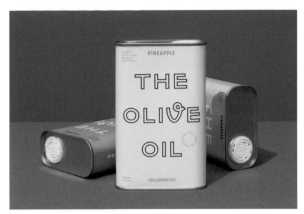

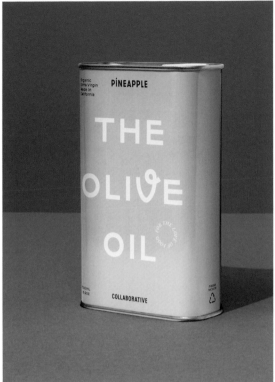

Dimensions:
98 mm × 155 mm × 44 mm

Material(s):
Tin, glass

Typeface(s):
Ginto, Viksjoe, Bell

Print & Finishing:
**Litho printing onto the tin plate
with gloss clear-cote**

● Pantone 7555 C	○ C: 0 M: 0 Y: 0 K: 0	● Pantone 295 C
C: 2 Y: 97 M: 28 K: 12		C: 100 Y: 0 M: 63 K: 67

● Pantone 2437 C	○ C: 0 M: 0 Y: 0 K: 0	● Pantone 295 C
C: 0 Y: 32 M: 24 K: 0		C: 100 Y: 0 M: 63 K: 67

○ C: 0 M: 0 Y: 0 K: 0	● Pantone 295 C	● Pantone 7555 C	● Pantone 2149 C
	C: 100 Y: 0 M: 63 K: 67	C: 2 Y: 97 M: 28 K: 12	C: 70 Y: 8 M: 33 K: 7

Gingibre

Design Agency: **Estúdio Bogotá**
Creative Direction: **Paula Cotta,**
Renata Polastri
Design: **Renata Polastri**
3D Design: **Helen Alves**
Client: **Gingibre - A Equilibrista**

According to Clovis de Barros Filho, 'happiness is an instant of life which is worth living all on its own.' It is also referred to as a 'power to act' according to Spinoza. Inspired by this abstract expressionism, Estúdio Bogotá worked on this concept using colour steam. The movement, energy and fusion of colours in a non-figurative representation bring us closer to the humane aspects of our own feelings. Gingibre's packaging uses three different fearless and bold colour combinations. The identity benefits from its own flexibility, unfolding into a system of parts intended to leave a mark and awaken users' desire for collectible items.

Dimensions:
123 mm × 56 mm (diametre)

Material(s):
**Aluminium can with
matte shrink sleeve**

Typeface(s):
Bely Display, Ofelia

Print & Finishing:
Flexography

C: 70	C: 70	C: 0
M: 0	M: 0	M: 60
Y: 20	Y: 45	Y: 70
K: 0	K: 0	K: 0
C: 25	C: 0	C: 0
M: 0	M: 85	M: 15
Y: 75	Y: 45	Y: 90
K: 0	K: 0	K: 0

Walker Brothers

Design Agency: **makebardo**
Creative Direction: **Bren Imboden, Luis Viale**

Founded in 2018, Walker Brothers brew traditional and high gravity (alcoholic) kombucha. The concept makebardo had for the brand was 'the bridge' because this symbol captures Walker Brothers' core values. They worked on a thin line between the literal and metaphoric. Instead of doing a pictorial bridge through an illustrative line work, makebardo decided to express the concept through the morphology of the wordmark. To close and reinforce the bridge concept, they created waves with movement that are strong and pure, representing a bold stream. At the same time, these waves make a perfect contrast with the subtle and elegant typography of the wordmark. Regarding the palette, makebardo opted for colours that balance freshness and maturity.

Dimensions:						
224 mm × 118 mm (diametre)						

Material(s):	●	○	●	○	●	○
Shrink sleeve label,	C: 0	C: 0	C: 80	C: 0	C: 0	C: 0
aluminum can	M: 45	M: 0	M: 45	M: 0	M: 80	M: 0
	Y: 100	Y: 0	Y: 80	Y: 0	Y: 100	Y: 0
	K: 0	K: 0	K: 45	K: 0	K: 0	K: 0

Typeface(s):				
BioRhyme, Belgium				

Print & Finishing:	●	○	●	○
Digital printing,	C: 100	C: 0	C: 50	C: 0
matte finishing	M: 70	M: 0	M: 90	M: 0
	Y: 40	Y: 0	Y: 35	Y: 0
	K: 30	K: 0	K: 15	K: 0

4Life Mineral Water
by Doi Chaang

Design Agency: **Prompt Design**

Springwater is naturally produced and is available from the source of mineral water from Doi Chaang, Chiang Rai (Northern Thailand). This water is the natural product from the abundant fertile forest which we have to respect for its habitat and environment. The package illustrations portray the clean blue lines representing the water and the adorable animals interacting with the lines, conveying how the animals live in water and reminding us of how water support all living things. In addition, the shadows of the animals and the motions of the ripples add an intriguing layer to the water bottle design subtly.

Dimensions:
220 mm × 60 mm (diametre)

Material(s):
PET plastic

Print & Finishing:
Gravure printing

C: 83	C: 0	C: 2	C: 0
M: 63	M: 35	M: 91	M: 0
Y: 0	Y: 85	Y: 83	Y: 0
K: 0	K: 0	K: 44	K: 94

C: 83	C: 20	C: 50	C: 0	C: 0
M: 63	M: 41	M: 70	M: 81	M: 100
Y: 0	Y: 66	Y: 80	Y: 100	Y: 100
K: 0	K: 1	K: 70	K: 0	K: 0

C: 83	C: 61	C: 66	C: 0
M: 63	M: 28	M: 34	M: 0
Y: 0	Y: 96	Y: 100	Y: 0
K: 0	K: 10	K: 19	K: 100

C: 83	C: 25	C: 35	C: 40	C: 40
M: 63	M: 25	M: 60	M: 65	M: 70
Y: 0	Y: 40	Y: 80	Y: 90	Y: 100
K: 0	K: 0	K: 25	K: 35	K: 50

C: 83	C: 17	C: 0
M: 63	M: 78	M: 0
Y: 0	Y: 100	Y: 0
K: 0	K: 6	K: 100

C: 83	C: 100	C: 100	C: 3	C: 15	C: 83
M: 63	M: 100	M: 46	M: 36	M: 100	M: 41
Y: 0	Y: 25	Y: 25	Y: 100	Y: 90	Y: 78
K: 0	K: 25	K: 25	K: 0	K: 10	K: 41

SEESAW WATER

Design Agency: **SEESAW Inc.**
Design: **Masahiro Sugawara, Yuko Saino, Yuka Chiba, Mahiro Kamioka, Seikyo Jo**

As part of SEESAW's internal project, a packaging design was created for the water bottles offered to visitors. Five of SEESAW's designers used illustrations and patterns to express their views and interpretations towards the theme, 'FUN × WAVE', which resulted in a design that shows the members' personalities.

Dimensions:
185 mm × 50 mm (diametre)

Typeface(s):
DIN Next LT Pro

Material(s):
PET bottle

Special Process:
Heat-shrinking

●	○
C: 0	C: 0
M: 0	M: 0
Y: 0	Y: 0
K: 100	K: 0

●	○	●
C: 0	C: 0	C: 0
M: 0	M: 0	M: 0
Y: 0	Y: 0	Y: 0
K: 15	K: 0	K: 100

●	●	●
C: 15	C: 75	C: 100
M: 100	M: 5	M: 90
Y: 85	Y: 100	Y: 40
K: 0	K: 0	K: 0
C: 70	C: 10	C: 25
M: 0	M: 15	M: 25
Y: 60	Y: 90	Y: 45
K: 0	K: 0	K: 0

●	
C: 90	C: 4
M: 69	M: 4
Y: 27	Y: 40
K: 16	K: 0
C: 51	C: 45
M: 55	M: 47
Y: 15	Y: 24
K: 0	K: 81

●	●	●
C: 55	C: 92	C: 80
M: 0	M: 74	M: 8
Y: 25	Y: 0	Y: 30
K: 5	K: 0	K: 10
C: 64	C: 5	C: 0
M: 13	M: 58	M: 0
Y: 0	Y: 80	Y: 0
K: 0	K: 0	K: 0

Izumi Odeki

Design Agency: **Mihara Minako Design**
Packaging Design: **Minako Mihara**
Graphic Design: **Koji Sato**
Calligraphy: **Yoshiko Nishimura**
Client: **Sibis Okumura**

Izumi Odeki is a craft's *sake* made in the sister cities of Izumi City and Katsuragi, located in the Kansai region of Japan. The calligraphy and the label were designed locally, and the wrapping paper features *ukiyo-e* artwork by the artist Utagawa Kuniyoshi. These particular sketches were chosen because of their playfulness and liveliness, which are kept in the spirit of the sake. The design team chose wrapping paper over boxes because it is lighter, uses less material and provides for a better presentation. The name 'Odeki' comes from a word commonly used during the Edo Period, which means 'very good', and Kuniyoshi wrote it next to the sketches he especially liked.

Dimensions:
Wrapping paper:
318 mm × 469 mm
Bottle: 750 ml

Material(s):
Paper

Typeface(s):
Handwritten calligraphy

Print & Finishing:
Offset printing

C: 78
M: 69
Y: 57
K: 18

C: 0
M: 70
Y: 90
K: 0

Modesta Cassinello

Design Agency: **Plácida, Buenaventura**
Illustration: **Buenaventura**
Photography: **Bruno Herrera**
Client: **Modesta Cassinello**

The new line of hair products by Modesta Cassinello
is inspired by the Mediterranean. Transmitting comfort
and a botanical essence, they aim at men and women
all around the world enjoying this kind of lifestyle.
Plácida wanted to avoid a static representation and
to create a visual system with different layers which
allow for including various new future products, sets
and merchandising. Therefore, they based their work
on a dynamic graphic system revolving around the
Mediterranean, which could adapt to different physical
and digital formats. Plácida focused on the scenery,
the land and the sea for an aesthetic result that is both
dynamic and colourful.

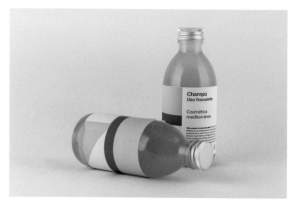

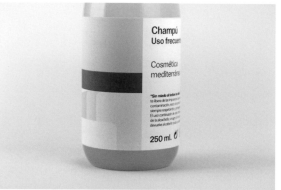

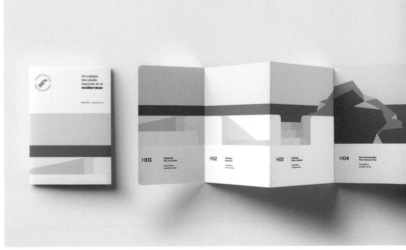

Dimensions:
135 mm × 65 mm (diametre)

Material(s):
**Ispira Purezza paper of
Fedrigoni, PET bottle**

Typeface(s):
Balto

Print & Finishing:
Offset printing

Pantone 660 C	Pantone 7457 C	Pantone 346 C	Pantone 149 C
C: 74 Y: 0	C: 25 Y: 0	C: 52 Y: 50	C: 0 Y: 56
M: 44 K: 0	M: 0 K: 0	M: 0 K: 0	M: 22 K: 0

Stand Your Ground

Art Direction: **Daria Kwon**
Design: **Daria Kwon**
Copywriting: **Yulia Tarasova**
Photography: **Daria Kwon**
Client: **Lintu**

'Stand Your Ground' is an unusual, practical and tasty gift—a set for making homemade liqueurs. In the box, consumers can find three sets of dry spices for making liqueurs: pepper-brandy, khrenovukha and Count Razumovsky's tincture. Spices for tinctures are already in half-litre bottles in required proportions. Add vodka, let it infuse and strain it off, then the drink is ready. The idea of the packaging design is to rethink the traditional Russian liqueur with a bold graphic language inspired by its ingredients.

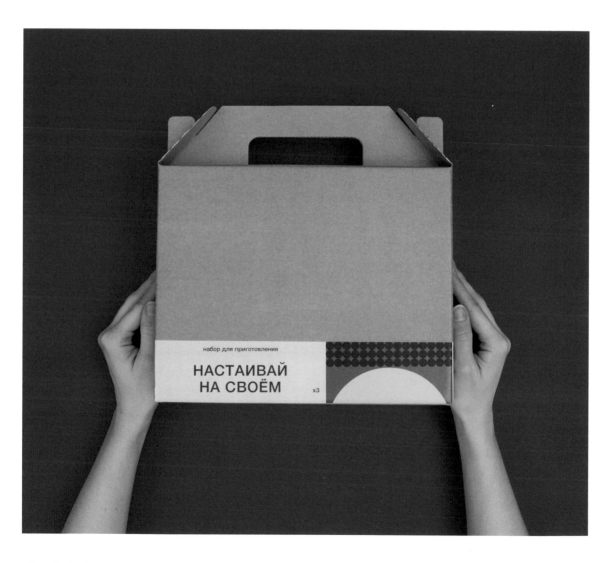

Dimensions:
Box: 260 mm × 260 mm × 90 mm
Bottle: 250 mm × 90 mm (diametre)

Material(s):
**Craft cardboard, glass,
coated paper**

Typeface(s):
Helvetica Neue

C: 30
M: 1
Y: 3
K: 0

C: 72
M: 81
Y: 6
K: 0

C: 22
M: 78
Y: 90
K: 0

C: 15
M: 13
Y: 62
K: 0

C: 24
M: 1
Y: 20
K: 0

C: 25
M: 97
Y: 79
K: 0

C: 15
M: 13
Y: 62
K: 0

C: 30
M: 1
Y: 3
K: 0

C: 77
M: 52
Y: 82
K: 13

C: 72
M: 81
Y: 6
K: 0

C: 24
M: 1
Y: 20
K: 0

C: 87
M: 47
Y: 71
K: 6

C: 77
M: 52
Y: 82
K: 13

C: 60
M: 96
Y: 65
K: 32

C: 1
M: 30
Y: 9
K: 0

C: 25
M: 97
Y: 79
K: 0

C: 22
M: 78
Y: 90
K: 0

Wine of Újpest

Design Agency: **Faraway Design**
Art Direction: **Bence Kovácsik**
Design: **Kristóf Balla**
Illustration: **Alíz Stocker**

Újpest Wine Week selects Wine of Újpest of the year annually, and the winner represents the district at official events for one year. Accordingly, Faraway Design designed a label for Újpest Municipality with the traditional characteristics of the district and the spirit of the winner on the bottles. A special feature of the concept is the crest shaped label—the left side will remain unchanged in the future while the right will feature an illustration inspired by the winning winery.

Dimension:
120 mm × 92 mm (diametre)

Material(s):
Paper, glass

Typeface(s):
AT Osmose

Print & Finishing:
**Digital printing, letterpress,
3D spot varnish**

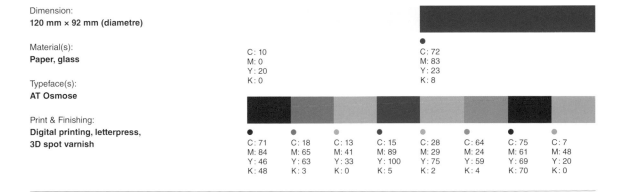

C: 10
M: 0
Y: 20
K: 0

C: 72
M: 83
Y: 23
K: 8

C: 71	C: 18	C: 13	C: 15	C: 28	C: 64	C: 75	C: 7
M: 84	M: 65	M: 41	M: 89	M: 29	M: 24	M: 61	M: 48
Y: 46	Y: 63	Y: 33	Y: 100	Y: 75	Y: 59	Y: 69	Y: 20
K: 48	K: 3	K: 0	K: 5	K: 2	K: 4	K: 70	K: 0

THE SAKERAKU

Design Agency: **WWAVE DESIGN**
Design: **Kenneth Ho, Ken Ho Ion Fat**
Photography: **Andrew Kan**

The project for the limited edition of Hirai Brewery's famous *sake* product 'Asajio'. Hirai is a sake brand founded more than 300 years ago in Otsu, Japan. The 18th successor and sake brewer teamed up with various designers and artists to create a distinctive sake package. For this global limited edition, WWAVE wanted to communicate its precious value through design and thus took inspiration from the famous local scenic spot—Lake Biwa. The sake labels are printed with different times and colour tones to express the unique scene of light shining on the lake at different time, connecting the value of 'Asajio' with the preciousness of time.

Dimensions:			
342 mm × 98 mm × 98 mm			

C: 59
M: 11
Y: 22
K: 0

C: 22
M: 22
Y: 56
K: 4

C: 74
M: 31
Y: 56
K: 17

Material(s):
White cardboard

C: 47
M: 44
Y: 7
K: 0

C: 11
M: 57
Y: 70
K: 0

C: 44
M: 65
Y: 52
K: 17

Typeface(s):
Custom type

Print & Finishing:
Hot-stamping with gold foil

C: 60
M: 30
Y: 4
K: 0

C: 16
M: 38
Y: 43
K: 3

C: 72
M: 51
Y: 32
K: 15

Lappa Olive - Premium & Organic Extra Virgin Olive Oil

Design Agency: **Ruto design studio**

The task was to create a chic and simple identity that would withstand the time and stays 'classic'. Ruto implemented the story of five virgins (a Roman story) to the concept and tried to blend it with the rich vegetation and the water springs of Lappa. To transfer the concept into the logo, they represented each virgin with a cavity at the end of the tree's roots. The symbol shows the origin of the highest quality olive oil. On the packaging, they used natural desaturated colours, with simple typography to create a vintage feeling depicting the long history of Lappa. Lastly, they made the cap from olive wood to achieve a more natural look.

Dimensions:
182 mm × 83 mm (diametre)

Material(s):
Glass, olive wood

Print & Finishing:
Silkscreen printing

C: 30
M: 0
Y: 22
K: 0

C: 0
M: 0
Y: 0
K: 100

C: 30
M: 0
Y: 22
K: 0

C: 0
M: 0
Y: 0
K: 0

PACKAGING - TAU WINZER TIESCHEN

Design Agency: **moodley design group**
Art Direction: **Daria Titarenko, Wolfgang Niederl**
Creative Direction: **Wolfgang Niederl**
Design: **Daria Titarenko, Adam Katyi**
Photography: **Stefan Robitsch,**
Manuel Schaffernak, Christian Niederl
Client: **Verein TAU Winzer Tieschen**

For the locals, Tieschen is not just a place in southeastern Styria, near the border with Slovenia. Rather, it is a way of life that connects them together. To lift the wine, classified as a DAC (Districtus Austriae Controllatus) local wine, into a new dimension and to make Tieschen a worthy wine ambassador, moodley design group used the framework of the DAC system for the design. The new branding exaggerates the idea by showing the name Tieschen as large as possible in analogue and digital form four times on the wine label. In addition, the diverse design variations of the new branding result in a pattern with a high recognition value which looks particularly good on the printed, stacked wine cartons.

Dimensions:
225 mm × 335 mm × 155 mm

Material(s):
Label, carton

Typeface(s):
**Le Murmure,
Sporting Grotesque,
GT America Condensed**

Print & Finishing:
**Letterpress, hot-stamping with
gold foil, offset printing**

Pantone Cool Gray 1 U

C : 0   C : 0
M : 0   M : 0
Y : 0   Y : 0
K : 100   K : 0

DIE-CUT PATTERNS

The die-cut templates can be downloaded and used for free for personal and non-commercial use only. If commercial use is involved, please contact the owner(s) for approvals. Sandu Publishing will not be responsible for any loss caused by unapproved commercial use.

Download link: http://www.sandupublishing.com/en/news/detail-52.html
Password: sanDuPUbLIshinG*2021

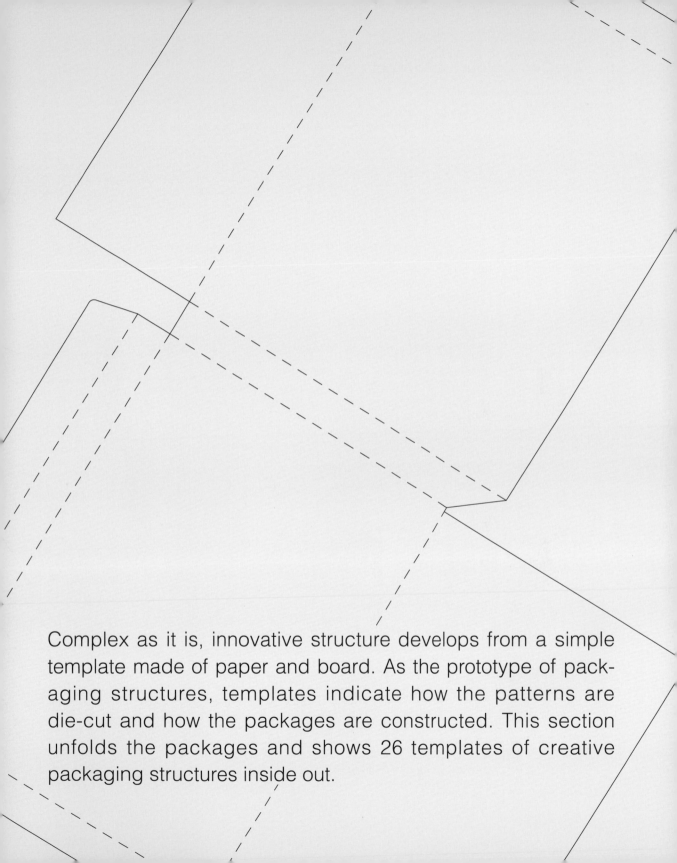

Complex as it is, innovative structure develops from a simple template made of paper and board. As the prototype of packaging structures, templates indicate how the patterns are die-cut and how the packages are constructed. This section unfolds the packages and shows 26 templates of creative packaging structures inside out.

01/ ZeeCoo Colourful Contact Lenses

Design Agency: **Hezilab**

Dimensions: **60 mm × 60 mm × 200 mm**
Material(s): **Cardboard**

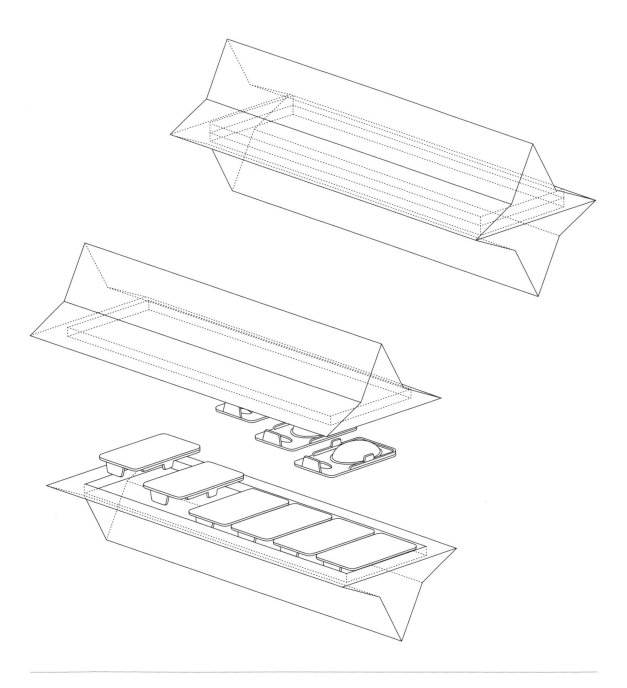

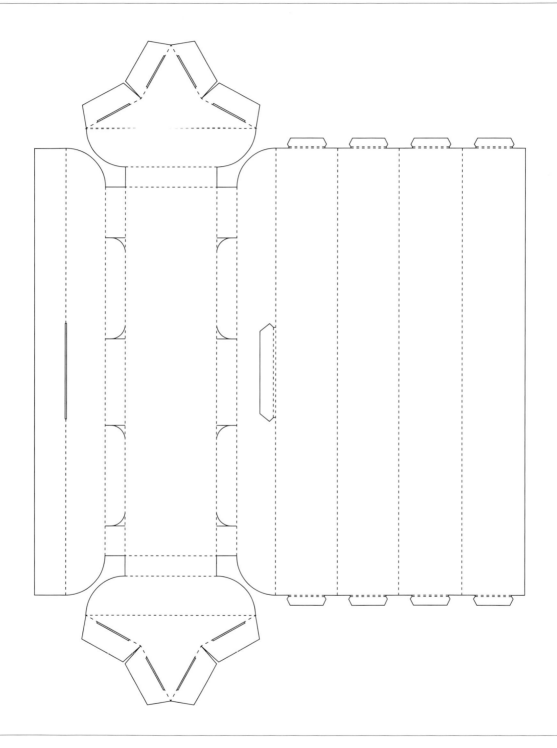

PALACE CERAMICS COLLECTION

Dimensions: **237 mm × 173 mm × 80 mm**
Material(s): **Card (3 mm thick)**

gluing

gluing

gluing

gluing

03/ **Fragments of Bournemouth**

Design: **Dominic Rushton**

Dimensions: **123 mm × 90 mm × 75 mm**
Material(s): **Greyboard (2 mm thick) with paper wrap (135 gsm)**

04/ **KK Love Item**

Design Agency: **nineteendesign**

Dimensions: **60 mm × 60 mm × 5 mm**
Material(s): **White carton (300 gsm)**

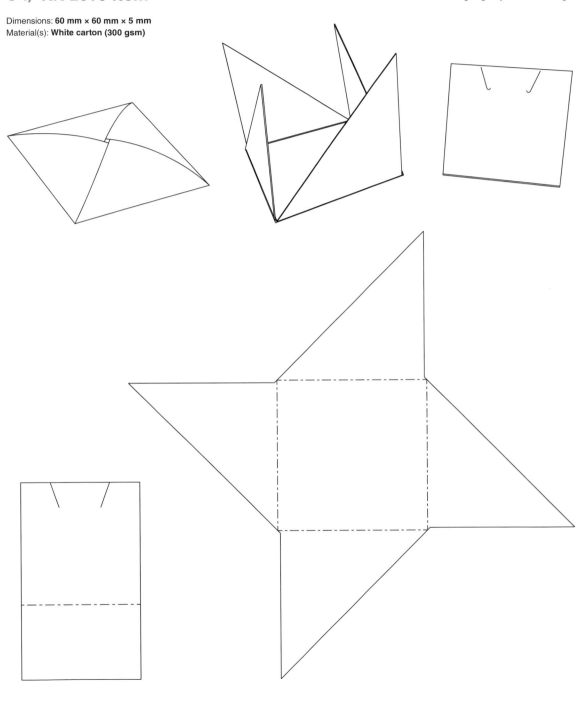

05/ Imanishi Socks

Design Agency: **Mihara Minako Design**

Dimensions: **119 mm × 125 mm × 73 mm**
Material(s): **Cardboard**

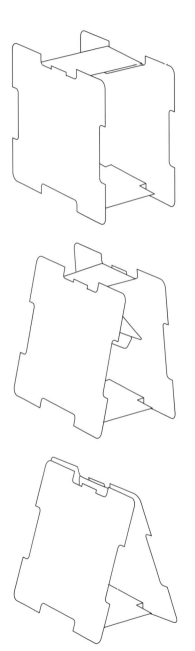

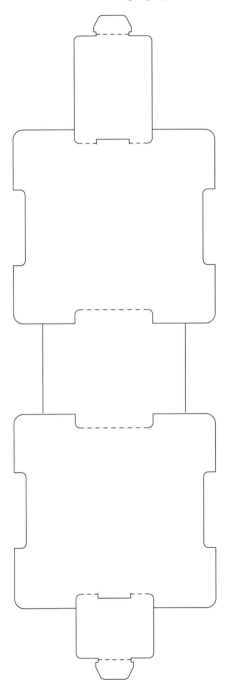

Design Agency: **nineteendesign**

Dimensions: **40.5 mm × 90.5 mm × 11.5 mm**
Material(s): **Black and white carton (300 gsm)**

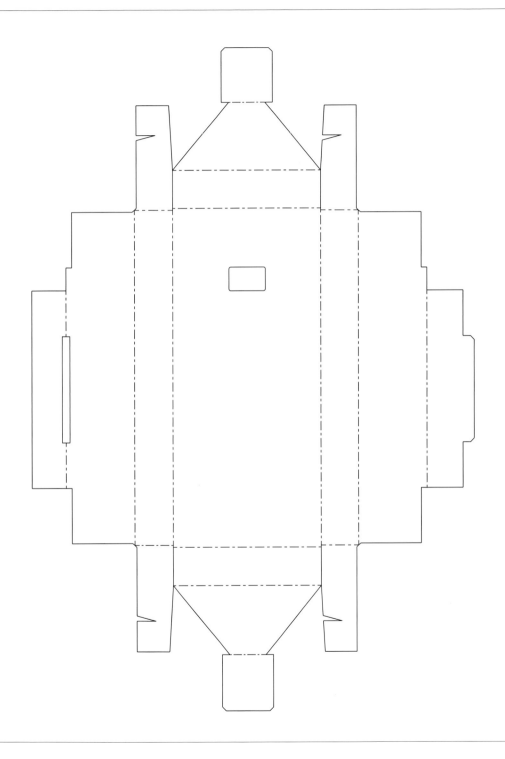

07/ CLAYD - MOUTHWASH MIST

Design Agency: **Plantis Inc.**

Dimensions: **98 mm × 38 mm × 33 mm**
Material(s): **Paper**

Dimensions: **135 mm × 65 mm × 65 mm**
Material(s): **Paper**

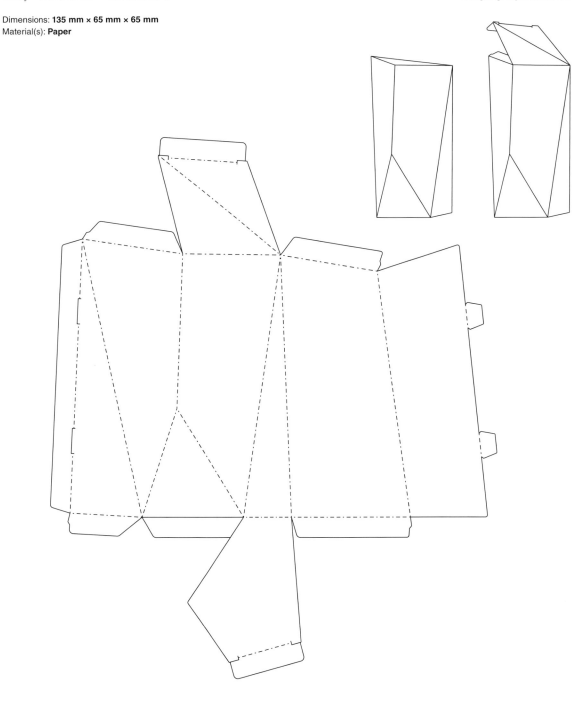

09/ CLAYD - CLAY MASK

Design Agency: **Plantis Inc.**

Dimensions: **148 mm × 70 mm × 53 mm**
Material(s): **Paper**

10/ Soulsister

Design: **Victoria Ng**

Dimensions: **106 mm × 122 mm × 45 mm**
Material(s): **Cardboard**

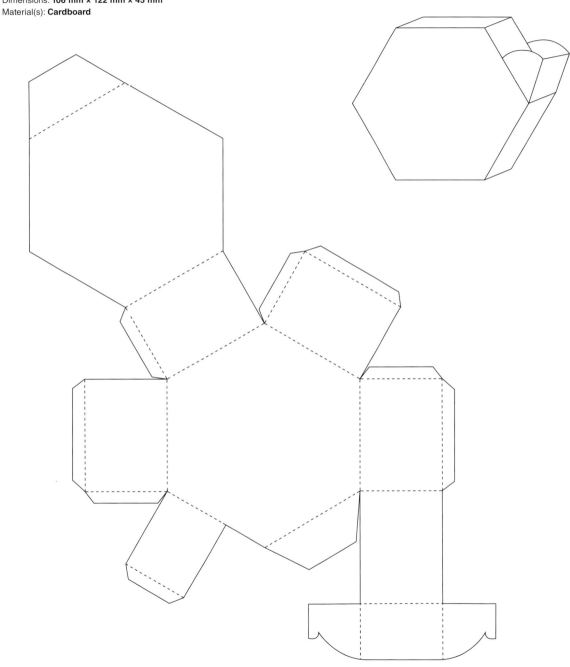

11/ DARLCHA TEA Branding & Packaging

Design Agency: AURG Design

Dimensions: **100 mm × 100 mm × 185 mm**
Material(s): **Carton**

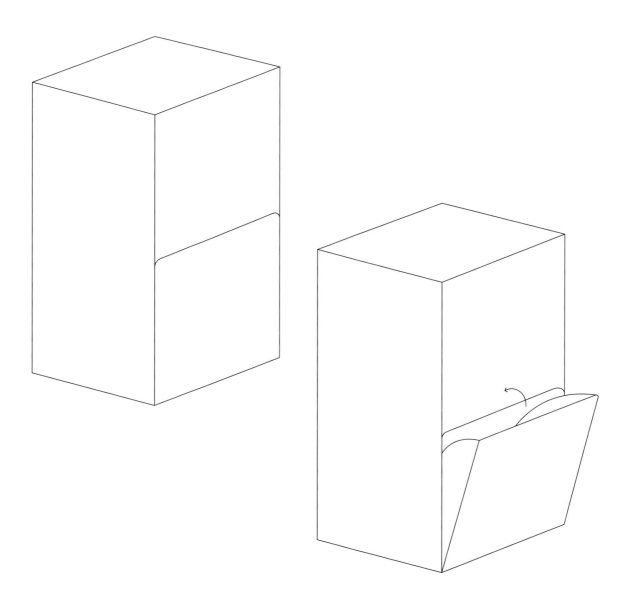

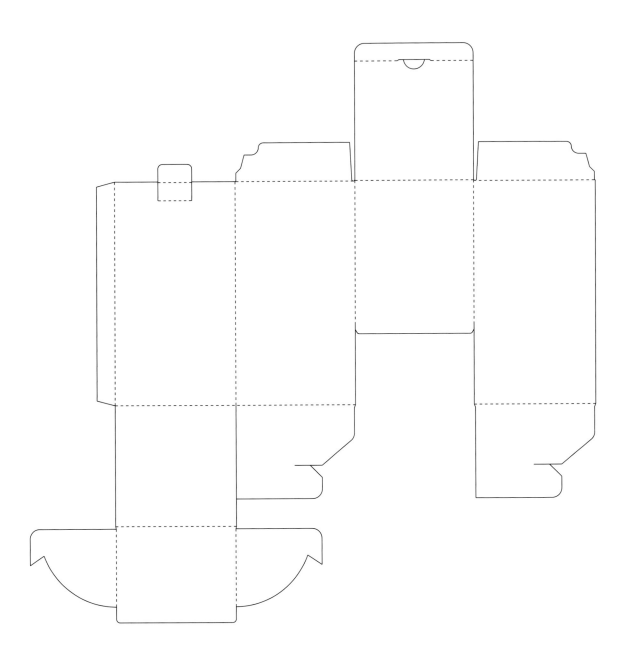

12/ **Brown Sugar Tea**

Design: **Li Shuk Yan, Fina**

Dimensions: **60 mm × 160 mm × 160 mm**
Material(s): **Paper**

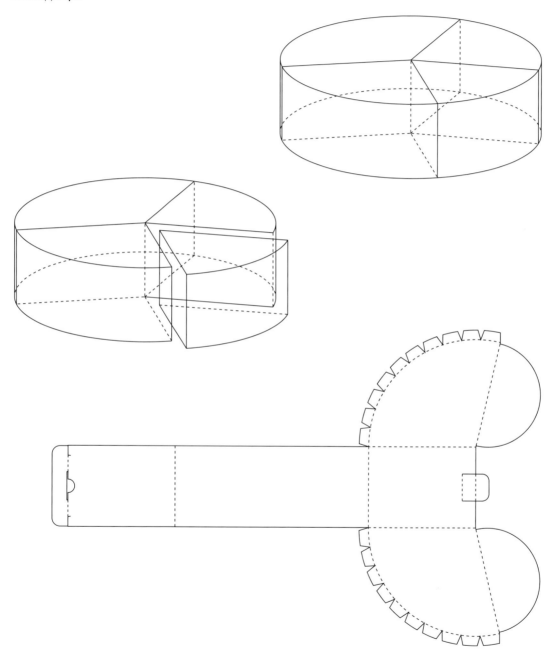

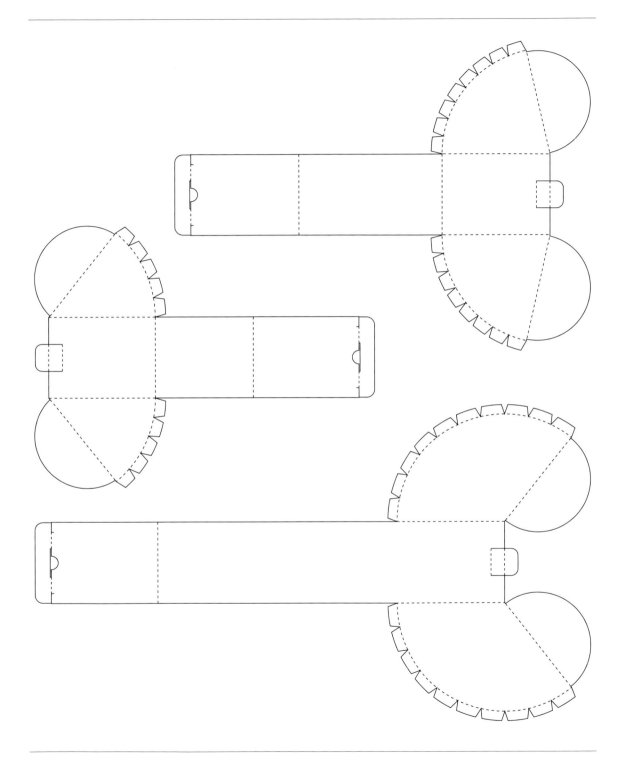

13/ FINE HERB

Design: **Yinan Hu**

Dimensions: **40 mm × 125 mm × 154 mm**
Material(s): **Cardboard**

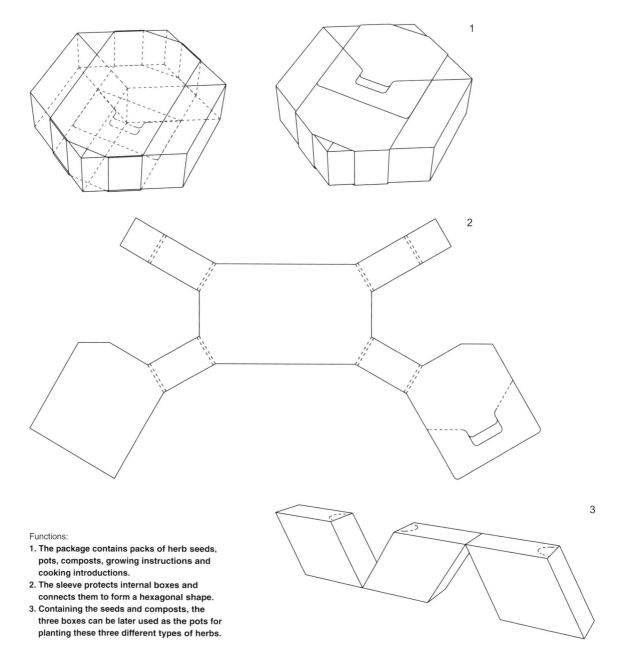

Functions:
1. The package contains packs of herb seeds, pots, composts, growing instructions and cooking introductions.
2. The sleeve protects internal boxes and connects them to form a hexagonal shape.
3. Containing the seeds and composts, the three boxes can be later used as the pots for planting these three different types of herbs.

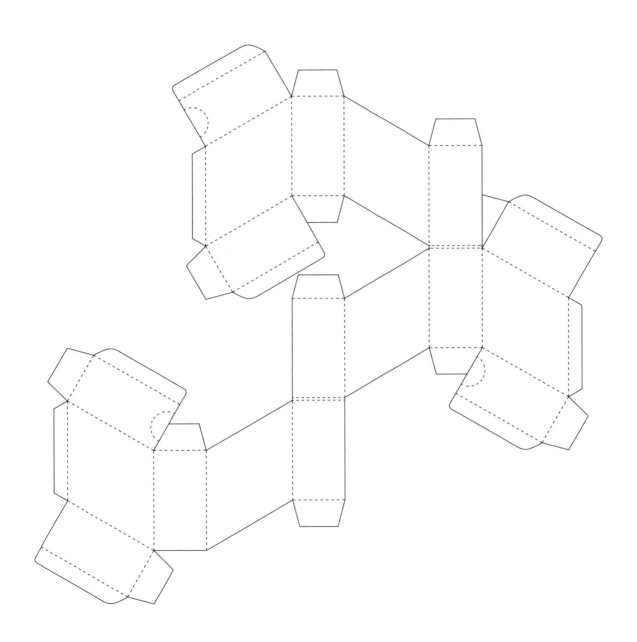

14/ Farma Pafylida Egg Packaging

Design Agency: **nineteendesign**

Dimensions: **183 mm × 146 mm × 65 mm**
Material(s): **Corrugated E'Flute with recycled carton (300 gsm)**

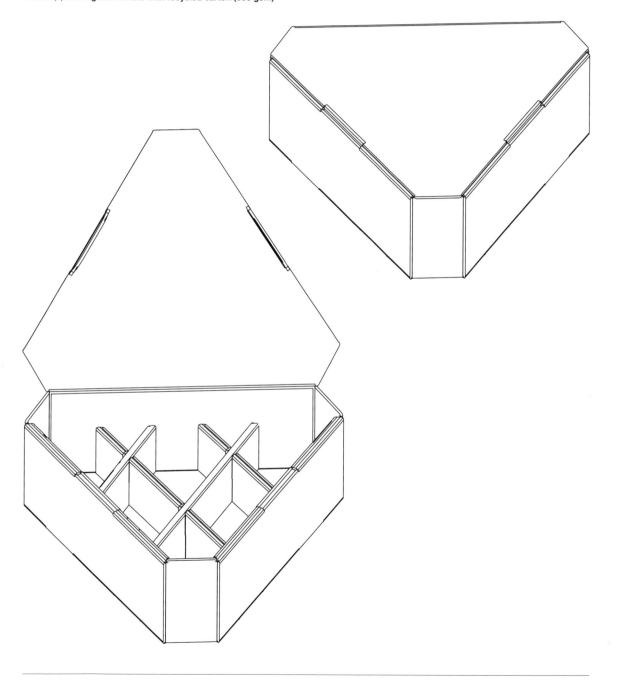

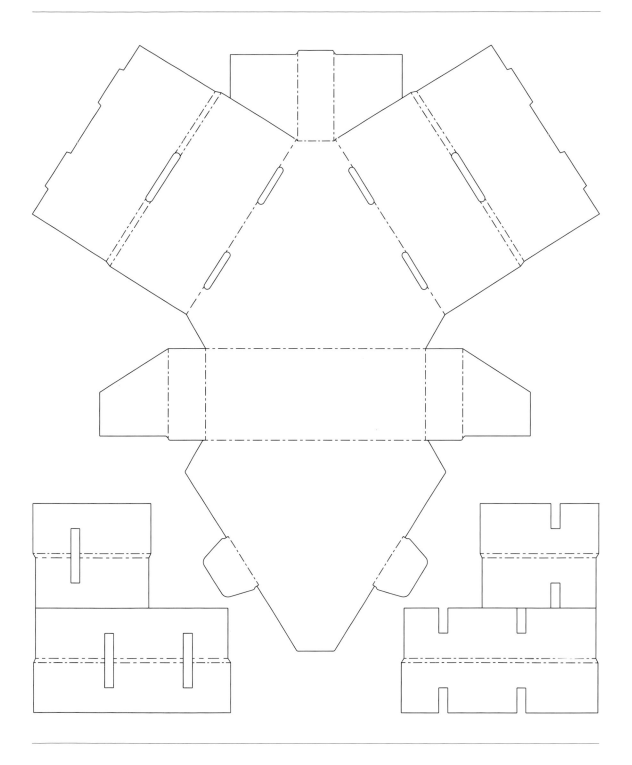

15/ KISSA Tea Company - Wagashi

Design: **Jane Wu**

Dimensions: **60 mm × 60 mm × 44 mm**
Material(s): **Matte paper stock (80 lb)**

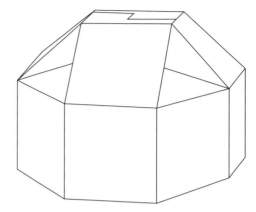

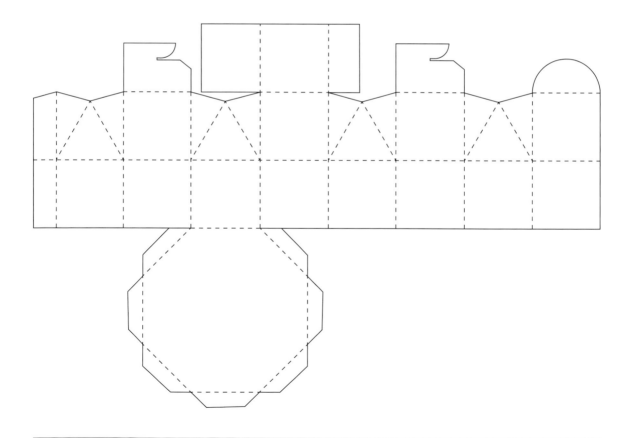

16/ **KISSA Tea Company - Loose Tea**

Design: **Jane Wu**

Dimensions: **270 mm × 270 mm × 189 mm**
Material(s): **Matte paper stock (80 lb)**

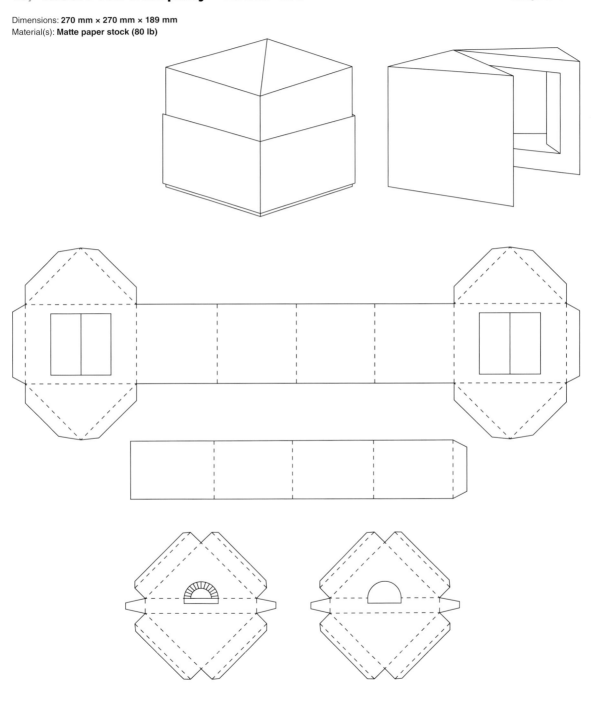

17/ Cocoon Box 1

Design: **Yolanda Sabuz, Sandra Solanas, Sofia Cuba**

Dimensions: **80 mm × 80 mm × 50 mm**
Material(s): **Cardboard**

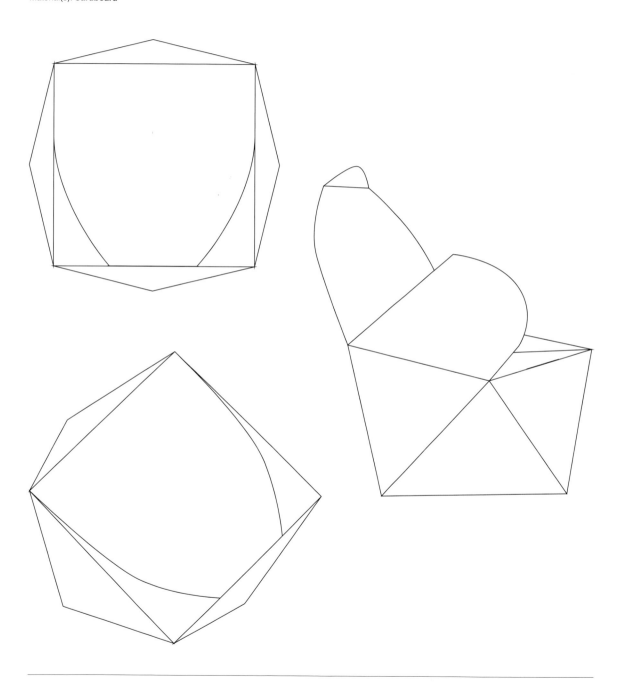

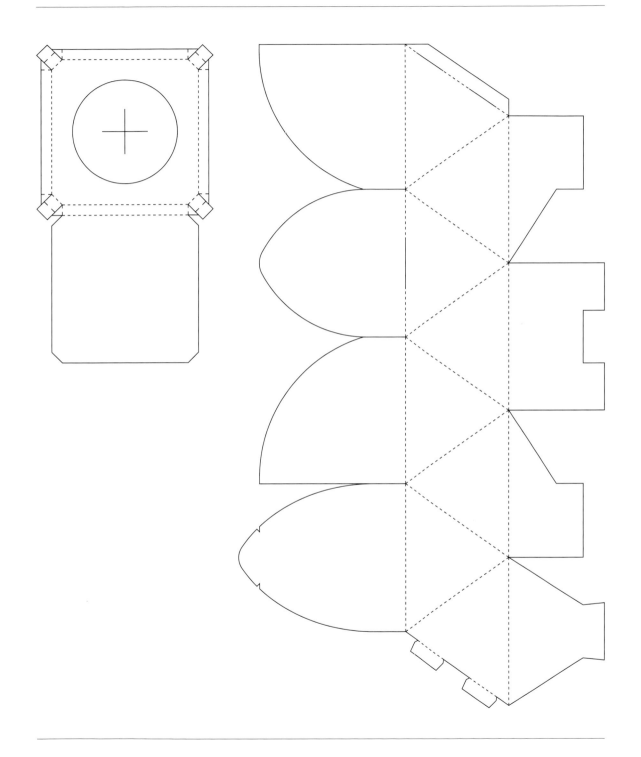

18/ Cocoon Box 2

Design: **Yolanda Sabuz, Sandra Solanas, Sofia Cuba**

Dimensions: **150 mm × 150 mm × 50 mm**
Material(s): **Cardboard**

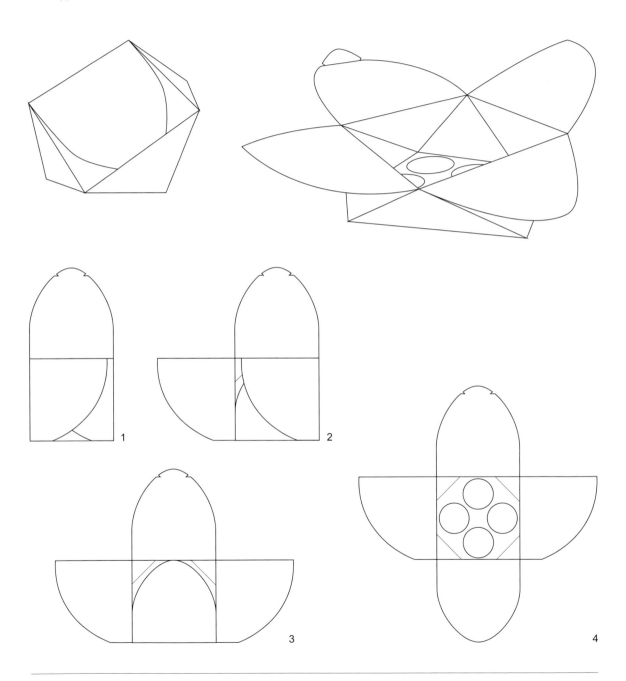

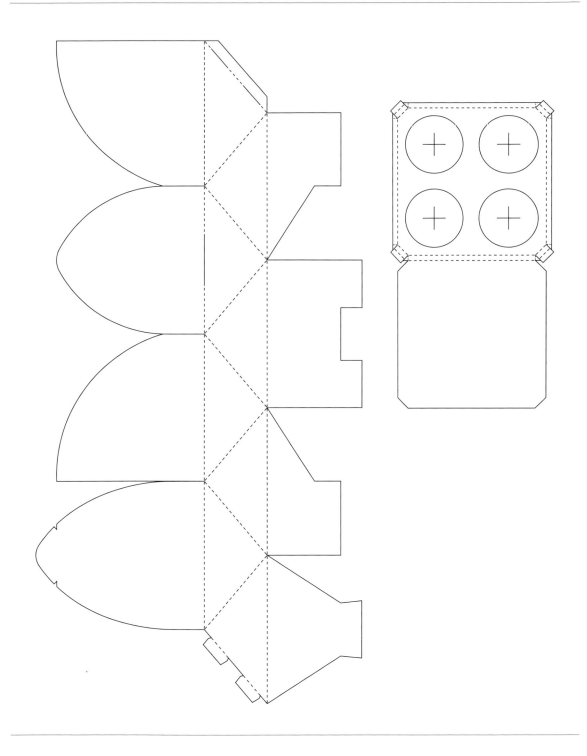

Design Agency: **AURG Design**

Dimensions: **70 mm × 70 mm × 90 mm**
Material(s): **Paper**

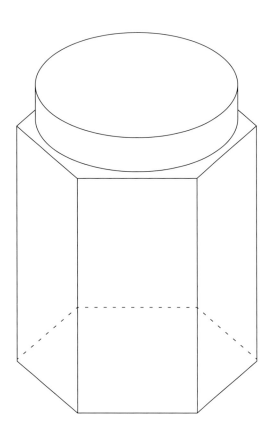

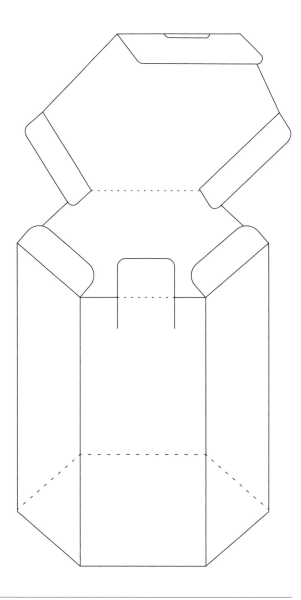

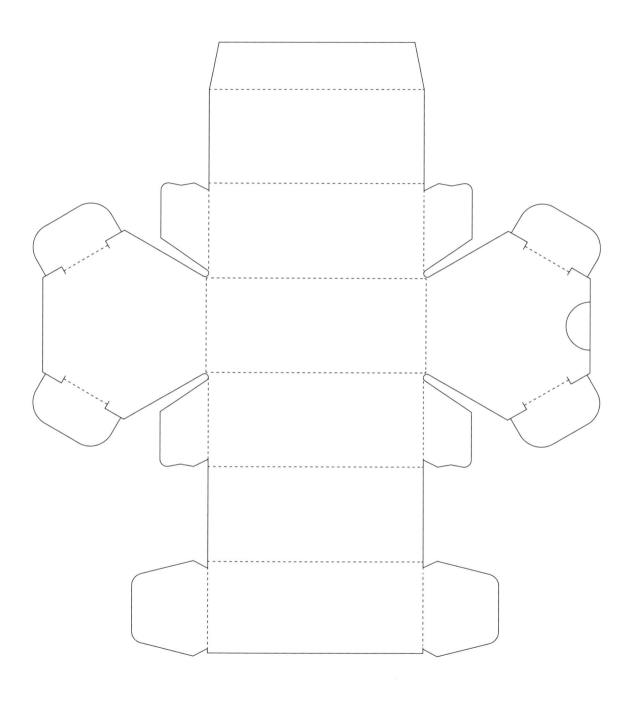

20/ **Grower Cosmetics**

Design: **Yan Zaretsky**

Dimensions: **300 mm × 100 mm × 100 mm**
Material(s): **Micro-cardboard, Favini Crush Olive Paper**

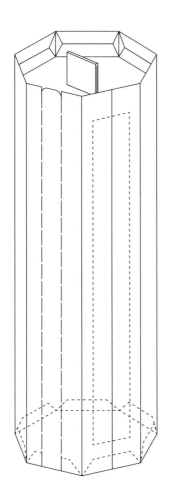

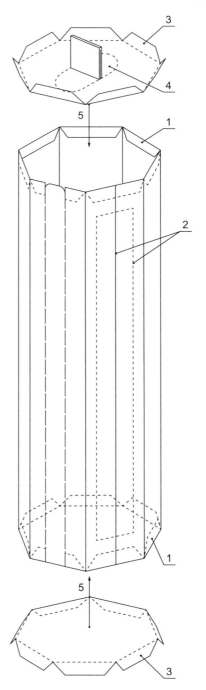

Assembly:
1. **Fold the bottom and top flaps and tape them to the body part.**
2. **Connect the opposite sides with a strip of taped cardboard.**
3. **Fold the flaps of the caps.**
4. **Tape the folded handle to the top cap.**
5. **Insert both caps into the body part.**

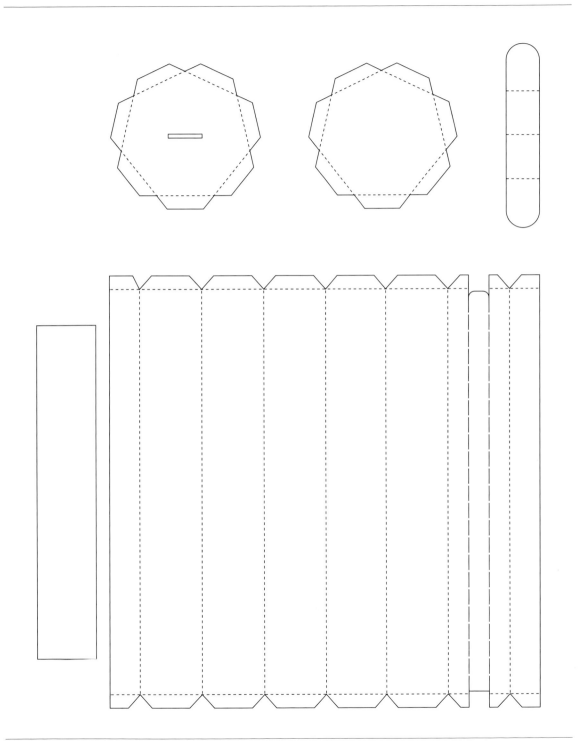

21/ Home Herby Home - Eco-packaging

Design: **Romina Alfonsi, Carla Estrada, Soha Savant**

Dimensions: **120 mm × 300 mm × 84 mm**
Material(s): **Cardboard**

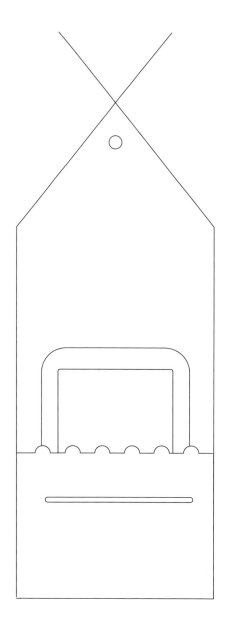

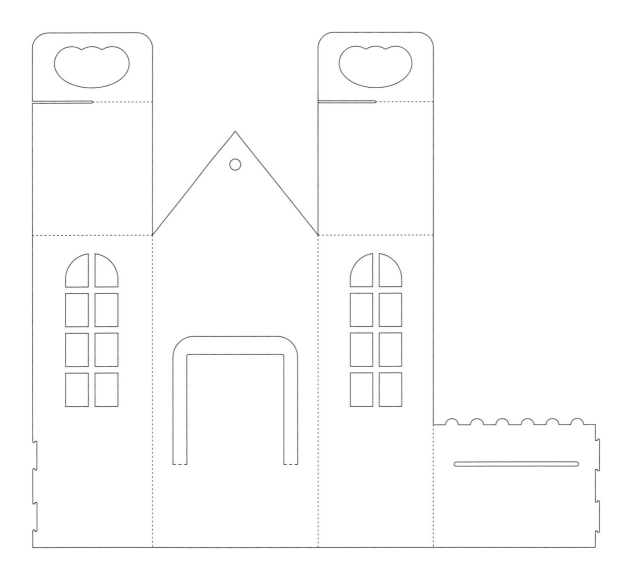

22/ Map of GBA Innovation and Entrepreneurship Town for Youth Talents

Design Agency: **SANDU Design**

Dimensions: **196 mm × 146 mm × 12 mm**
Material(s): **CREATIVE BOARD ROYAL (270 gsm)**

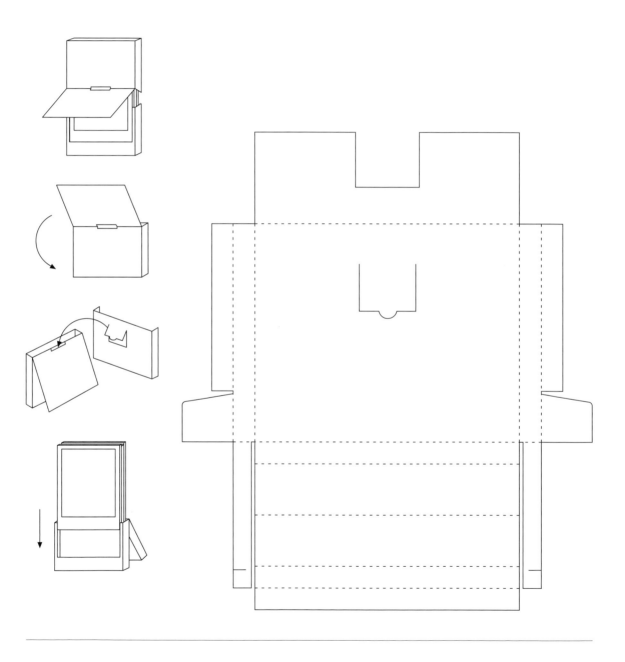

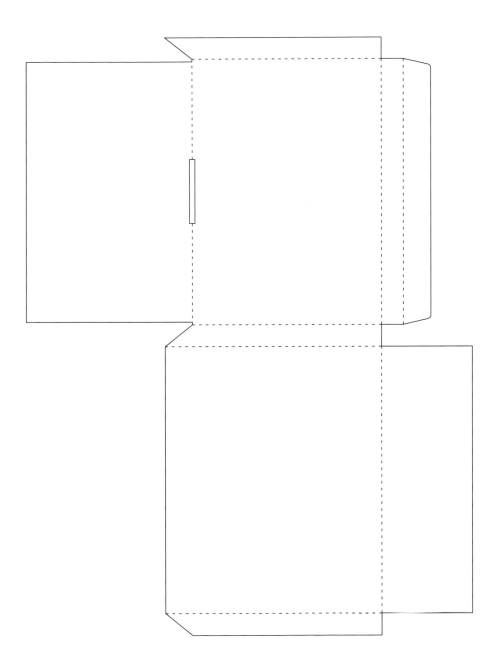

23/ MAD Fashion - Packaging

Design Agency: **Forward Design Studio**

Dimensions: **50 mm × 130 mm × 130 mm**
Material(s): **White cardboard (350 gsm), Duplex (400 gsm)**

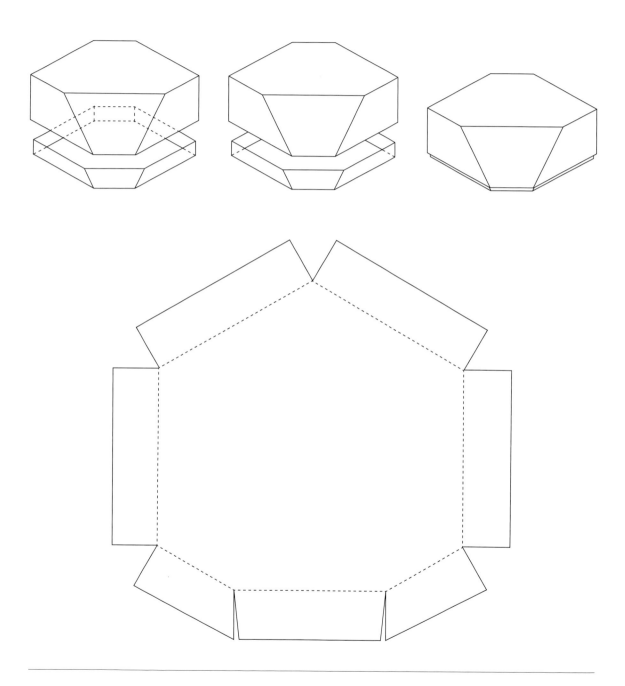

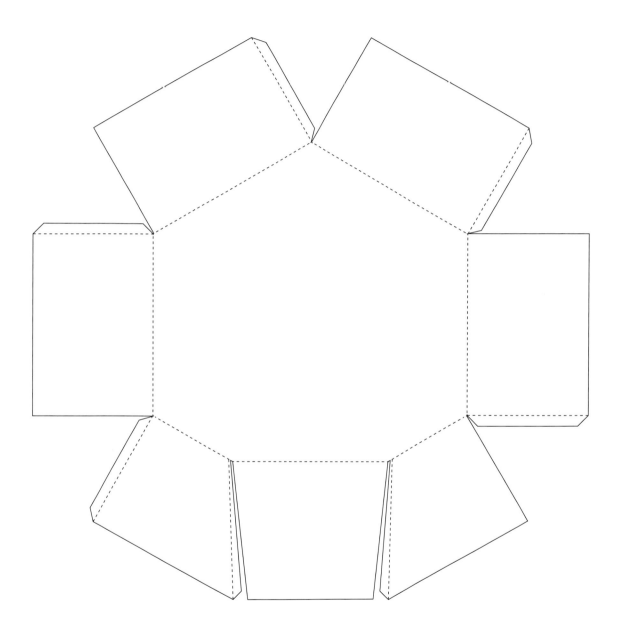

24/ **Hay Glass**

Design: **Xiaoxiao Ma**

Dimensions: **150 mm × 216 mm × 218 mm**
Material(s): **E-flutes Cardboard**

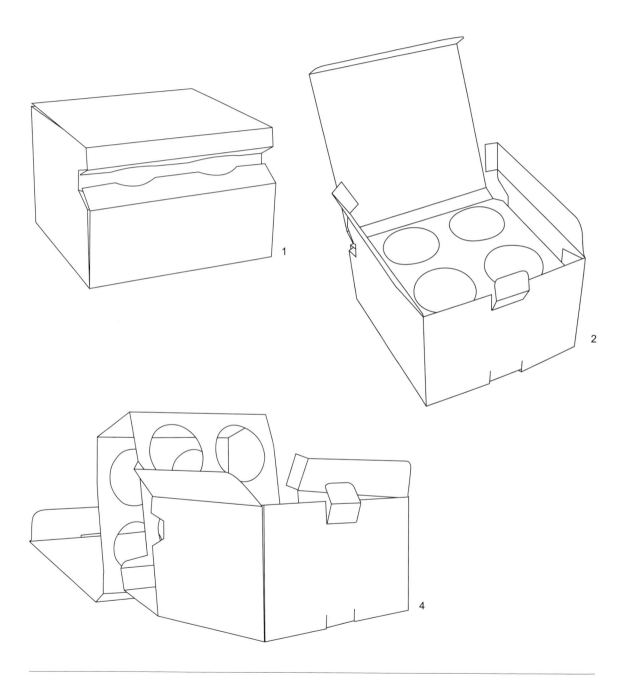

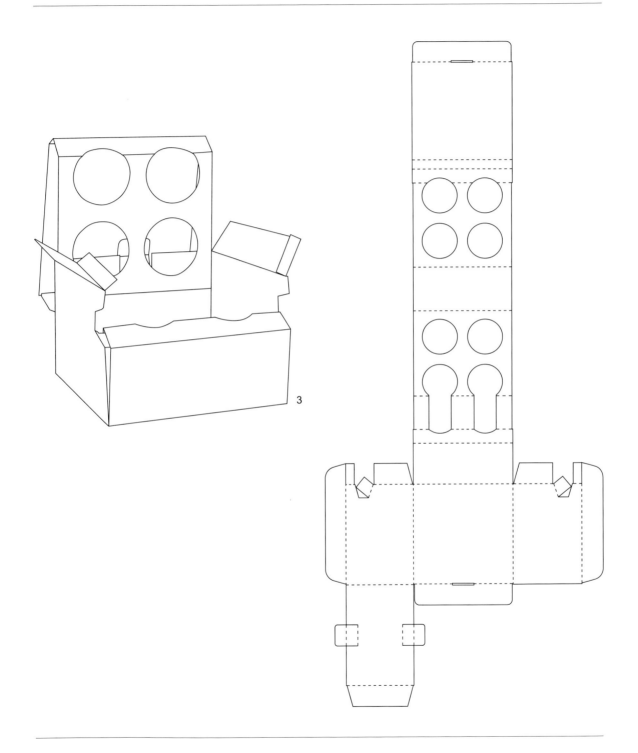

3

25/ On the Tip of a Tongue

Design: **Diana Molyte, Dovile Kacerauskaite**

Dimensions: **180 mm × 180 mm × 260 mm**
Material(s): **Paper**

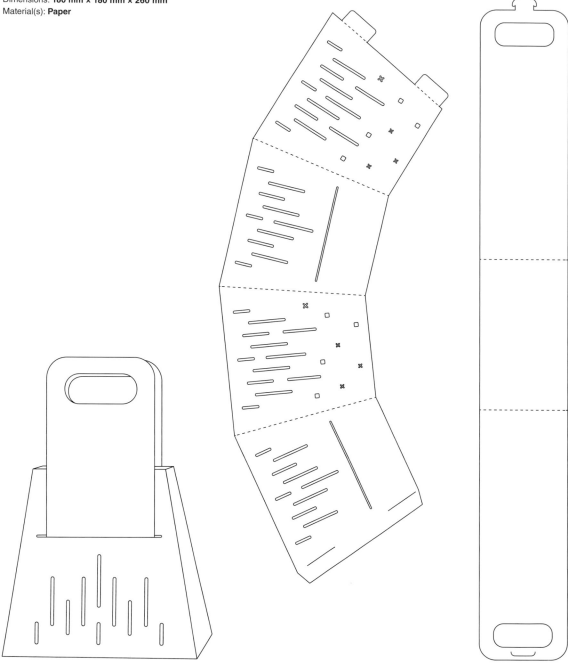

26/ POCK BOWL

Design: **MEETON**

Dimensions: **270 mm × 92 mm × 198 mm**
Material(s): **Cardboard**

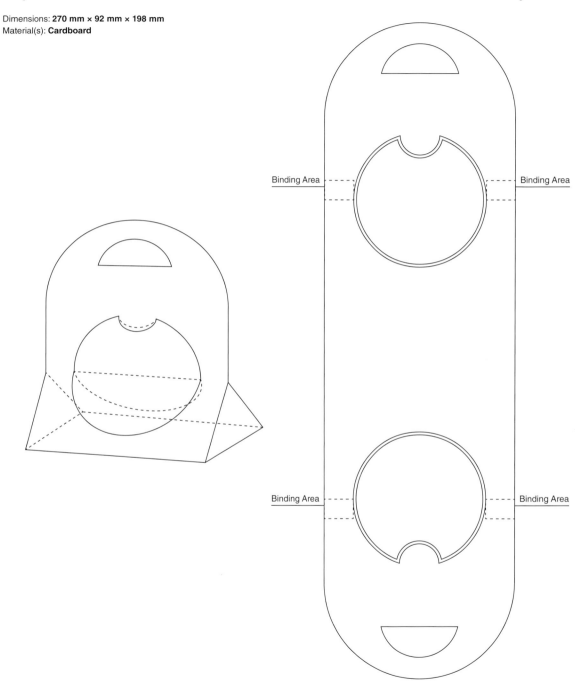

Binding Area

Binding Area

Binding Area

Binding Area

Credits

01/ ZeeCoo Colourful Contact Lenses

Design Agency: **Hezilab**
Art Direction: **Luo Haozhen**

02/ PALACE CERAMICS COLLECTION

Design Agency: **PMDESIGN**
Design: **Paulo Marcelo**
Packaging production: **AÉME premium packaging**
Photography: **Álvaro Martino**
Client: **Atelier Daciano da Costa**

03/ Fragments of Bournemouth

Design: **Dominic Rushton**
Institution: **Arts University Bournemouth**

04/ KK Love Item

Design Agency: **nineteendesign**
Client: **KK Jewelry Lab**

05/ Imanishi Socks

Design Agency: **Mihara Minako Design**
Design: **Minako Mihara**
Die-Cutting: **Nakanokigata Inc.**
Client: **Imanishi Socks**

06/ KK Lucky Charm 2019

Design Agency: **nineteendesign**
Client: **KK Jewelry Lab**

07/ CLAYD - MOUTHWASH MIST

Design Agency: **Plantis Inc.**
Design: **Keijiro Yamaguchi**
Client: **Mother Earth Solutions Co., Ltd**

08/ CLAYD - SERUM

Design Agency: **Plantis Inc.**
Design: **Keijiro Yamaguchi**
Client: **Mother Earth Solutions Co., Ltd**

09/ CLAYD - CLAY MASK

Design Agency: **Plantis Inc.**

Design: **Keijiro Yamaguchi**
Client: **Mother Earth Solutions Co., Ltd**

10/ SoulSister

Design: **Victoria Ng**

11/ DARLCHA TEA Branding & Packaging

Design Agency: **AURG Design**
Client: **Darlcha Company**

12/ Brown Sugar Tea

Design: **Li Shuk Yan, Fina**
Photography: **Li Shuk Yan, Fina**

13/ FINE HERB

Design: **Yinan Hu**
Photography: **Fei Wang**

14/ Farma Pafylida Egg Packaging

Design Agency: **nineteendesign**
Client: **Pafylida Farm**

15/ KISSA Tea Company - Wagashi

Design: **Jane Wu**
Photography: **Jane Wu**

16/ KISSA Tea Company - Loose Tea

Design: **Jane Wu**
Photography: **Jane Wu**

17/ Cocoon Box 1

Design: **Yolanda Sabuz, Sandra Solanas, Sofia Cuba**
Supervision: **Emili Padrós, Ion Elola**
Institution: **Elisava Barcelona School of Design and Engineering**

18/ Cocoon Box 2

Design: **Yolanda Sabuz, Sandra Solanas, Sofia Cuba**
Supervision: **Emili Padrós, Ion Elola**

Institution: **Elisava Barcelona School of Design and Engineering**

19/ CHAEON

Design Agency: **AURG Design**
Design: **Songyee Paek**

20/ Grower Cosmetics

Art Direction: **Yan Zaretsky**
Design: **Yan Zaretsky**
Print Production: **Alan Bur**
Photography: **Anatoly Vasiliev**
Client: **Grower Cosmetics**

21/ Home Herby Home - Eco-Packaging

Design: **Romina Alfonsi, Carla Estrada, Soha Savant**
Photography: **Romina Alfonsi, Carla Estrada, Soha Savant**

22/ Map of GBA Innovation and Entrepreneurship Town for Youth Talents

Design Agency: **SANDU Design**
Design: **None, Human Mar, Huang Haishu, Zikho**

23/ MAD Fashion - Packaging

Design Agency: **Forward Design Studio**
3D: **Gustavo Lacerda**

24/ Hay Glass

Design: **Xiaoxiao Ma**
Supervision: **Thomas McNulty**
Photography: **Xiaoxiao Ma**

25/ On The Tip of A Tongue

Design: **Diana Molyte, Dovile Kacerauskaite**

26/ POCK BOWL

Design Agency: **MEETON**
Design: **Julia Cao**
Photography: **Shawn Huang**

INDEX

BY-ENJOY DESIGN

by-enjoy.com

BY-ENJOY is an innovative design studio that uses the creative thinking of 're-understanding and then establishing communication' to carry out multi-media and cross-project visual presentations in the field of culture and branding.

P042–043

Choice Studio

choice.studio

Choice Studio focuses on brand design and offers sincere solutions under the source of creativity.

P068–069

Chris Trivizas Design

christrivizas.gr

Chris Trivizas Design is a creative studio committed to providing premium quality and integrated communication solutions tailored to each client's needs. Their services include branding, corporate identity, logotype and packaging.

P142–143

Daria Kwon

behance.net/DariaKwon

Daria Kwon is an art director, graphic and web designer, letterer and stylist-photographer.

P260–261

Diana Molyte, Dovile Kacerauskaite

diamol.graphics
behance.net/DovileKacerauskaite

Diana Molyte and Dovile Kacerauskaite are young designers who appreciate minimalism and search for innovative and courageous decisions when it comes to packaging, branding and illustrations projects.

P310

Dominic Rushton

instagram.com/domrushtondesign

Dominic Rushton has shown an interest in design from an early age and now focuses on packaging and branding design as a career.

P275

Duy Dao

duydao.net

Duy Dao is an award-winning designer and art director whose work has been recognised and published by numerous leading organisations and awards.

P122–123

Echo Yang

echoweilunyang.com

Echo Yang is a freelance graphic designer who is fascinated by printed matter and graphic expression. She completed her master's degree in Information Design from Design Academy Eindhoven.

P210

Estúdio Bogotá

estudiobogota.com.br

Estúdio Bogotá is a design office established by Paula Cotta and Renata Polastri. Their attention to details goes throughout the creative process of graphics that translate these decisions with quality, efficiency and technical excellence.

P248–249

Evelina Baniulyte, Gabija Platukyte

behance.net/evelinabaniulyte
behance.net/gplatukyte

Evelina Baniulyte and Gabija Platukyte met during their school days and both fell in love with carefully selected details in the design process.

P032–033

F33

fundacion33.com

F33 is a creative agency based in Murcia, Spain, since 2006. They look for creative solutions developed by a conceptual base and adequate graphic support.

P172–173

Faraway Design

faraway.hu

Faraway Design is a design agency. A group of copywriters, graphic, product, motion and web designers who believe design is not just about nice typefaces and graphics but the harmony of words and shapes, the connection of meaning and function.

P262–263

Forward Design Studio

behance.net/forwarddesignstudio

Forward is a Brazilian design studio specialised in brand identity and packaging design.

P306–307

Frame inc.

frame-d.jp

Frame inc. is a Japanese countryside design office aiming for production that will compete with the world even from the countryside.

P204–205

Fulya Kuzu

behance.net/fulyakuzuu

Fulya Kuzu is a young visual communication designer exploring her strength and passion in creating corporate identities, packaging designs, editorial designs and photo manipulations of any kind.

P134–135

GM Creative

behance.net/gaomuoicreative

'GM' stands for 'Gao Muoi' in Vietnamese, which is the hope to make ends meet, delivering the Viet heart and spirit through visual communication.

P056–057, 198–199

GRECO Design

grecodesign.com.br

GRECO is a graphic design consultancy with a focus on visual identity, editorial, package and wayfinding. They believe that appealing to their senses, design makes clear the differences between brands and makes the objectives real.

P150–151

Hansen/2

hansen2.de

Hansen/2 is a full-service design studio in Hamburg, Germany. They develop identities through clear design solutions, focusing on brands, strategy and design.

P128–129

Herefor Studio

hereforstudio.com

Herefor is a branding and packaging design studio in Brooklyn and Honolulu, America. They make nice things for nice people, nicely.

P244–245

Hezilab

hezilab.com

Hezilab is an exquisite and enthusiastic design team delving into product and packaging design and persisting in exploration. They think and try boldly and long to work with clients to develop surprising products together.

P272–273

HOOOLY DESIGN

hooolydesign.com

HOOOLY is a creative team with the core of design and visual arts, aiming to drive the brands to create unique experience through design. They have launched a series of independent design projects to communicate with the young.

P120–121

Hsu Shih-Chien

behance.net/roy8787roy1e4b

Hsu Shih-Chien is an image creator who is good at commercial design, film-making, motion design and animation. Recently, he is focusing on digital animation.

P136–137, 160–161

Hunk Xing

behance.net/ihunk

Graphic designer Hunk Xing has been invited to the recording of the BBC and CCTV's documentary *China's Greatest Treasures* and creative live broadcast of Adobe Live. His work has won Design for Asia Awards Silver Award, Graphic Design in China Award, Award 360° 100 Design of the Year and so on.

P226–227

INPIN DESIGN

inpindesign.com

Under the core concept of seeing the truth, goodness and beauty and connecting the people, event and things, INPIN DESIGN's goal is to combine their design ideas and marketing integration capabilities to create solutions and value.

P040–041

Ivan Gvozdev

behance.net/ivan_gvozdev

Ivan Gvozdev is a graphic designer who has been involved in branding, packaging design and industrial design since 2014. From 2018 to the present, he has been working in the Weiss Water branding agency.

P036–037

Jane Wu

letjanedesign.com

Jane Wu is an interdisciplinary designer who graduated from Pratt Institute and obtained a Master of Science degree in Package Design with distinction.

P292–293

Javier Garduño Estudio de Diseño

javiergarduno.com

Javier Garduño founded his namesake studio in 2010. He combines his 20 years of expertise with a fresh and new vision of graphic design. He and his members achieve their results through a multidisciplinary approach.

P070–071

Joanna Shuen

joannashuen.com

From print to digital, Joanna Shuen's practice involves playful experimentation and research that combine analogue processes with digital technology. The result is a unique visual style with strong, well-informed conceptual solutions.

P130–131

K9 Design

k9designstudio.com

K9 Design expects themselves to have strong sensitivity for design and aesthetics. They believe that communication will lead to good creation and work hand-in-hand with their clients to produce quality, valuable and trendy works.

P016–017, 026–027, 076–077, 242–243

Karla Heredia Martínez

karlaheredia.com

Karla Heredia Martínez has travelled around the world and worked with people of different cultures to learn, respect and value the differences and thus builds her worldview and becomes a good and human designer.

P200–201

Kenneth Kuh

kennethkuh.info

As a designer, Kenneth Kuh strives for opportunities to create a more connected world. Through art and design, he wants to become a storyteller who curates stories drawing intersections between people.

P072–073

Kim Hyun-Tae, Son Yoon-Joo

behance.net/stu2809c9e8
behance.net/YunJuSon

Kim Hyun-Tae and Son Yoon-Joo are designers based in Korea.

P146–147

Kingdom & Sparrow

kingdomandsparrow.co.uk

Kingdom & Sparrow is an agency for a

new generation of brands. Based in their studio by the sea in Falmouth, England, their team combines fresh thinking and imaginative design to elevate and transform the brands of today.

KITADA DESIGN Inc.

kitada-design.com

KITADA DESIGN Inc. was established by Shingo Kitada in 2015.

Li Shuk Yan, Fina

behance.net/finali

Li Shuk Yan, Fina, is specialising in branding, packaging and illustration. She aims to create a good interaction between her design and the audience as she believes design should communicate with people to enhance their daily life.

LOCO Studio

l-o-c-o.com

LOCO is a branding agency providing services in branding, graphic design, packaging, design of periodical literature. Also, they create ideas for public events, make up the designs and later organise and implement them into real life.

low key Design Company

behance.net/lowkeydesigncompany

low key Design is an independent design studio based in Shanghai. Their graphics open the door for brands to connect with people.

makebardo

makebardo.com

The creative design studio makebardo is run by Bren Imboden and Luis Viale. They have chosen to keep the studio small, allowing them to work closely with their clients, giving them personalised attention without losing the quality of the work.

Marco Arroyo-Vázquez

behance.net/marcoarroyevazquez95

Marco Arroyo-Vázquez is a packaging lover whose philosophy of work is 'work hard, design harder'.

MARINA GOÑI STUDIO

marinagoni.com

MARINA GOÑI is a branding studio committed to design, service quality and their clients. Each project is an opportunity for them to learn about branding, strategy and design, the team and working processes.

Martin Naumann, Andrius Martinaitis

behance.net/mnaumanndesign
behance.net/andriusamam

Martin Naumann is a digital artist and graphic designer specialising in abstract generative art.

Andrius Martinaitis is a freelance package designer and a chocolatier himself.

Masahiro Minami Design

masahiro-minami.com

Located in Shiga, Japan, Masahiro Minami Design is good at designing for various local industries and dealing with diverse materials in design. Their main works include graphics, packages, products and branding.

Máté Knapecz

behance.net/mateknapecz

Máté Knapecz, feeling interested in several forms of graphic design, is currently studying at a university in Hungary.

Mechi Co. Design

mechicodesign.com

Mechi Co. Design is a duo collective that focuses on branding, packaging and illustration design. They are curious, searching for different approaches and new combinations. They transform intangible concepts into real experiences that share a story.

MEETON

behance.net/meeton

MEETON is a design studio seeking creativity and originality in a thoughtful visual language.

Michelle Currie

michellecurriedesign.com

Through colourful palettes and bold visual elements, Michelle Currie creates functional and visually appealing designs by combining her passion for the fine arts and illustration and love for organisation and order.

Mihara Minako Design

behance.net/miharaminako

Minako Mihara is a Japanese package designer who established her namesake design studio. She is specialised in packaging design and offers cost estimation and marketing strategies for designing boxes and structures with varying types of materials.

milieu studio

milieu-studio.com

milieu studio gathers a group of young design professionals around the world, hoping to understand their clients' stories and come up with sustainable and innovative solutions.

Minsoo Kang, Jeongwon Kim

behance.net/riverpushwater
behance.net/0508kjwe0cd

Minsoo Kang and Jeongwon Kim are graphic and product designers based in Seoul, Korea.

P058–059

moodley design group

moodley.at

moodley design group is an internationally active strategic brand and design company, specialising in designing future-oriented brands, products and services. The members join together to form interdisciplinary and highly-focused teams for different projects.

P268–269

nineteendesign

nineteendesign.gr

nineteendesign believes that the optimum result is achieved through the collaboration of designers specialised in different fields, who share the same design vision. This ensures the continuation of a successful design methodology.

P276, 278–279, 290–291

nomo®creative

nomocreative.com

nomo®creative is a design studio with different experienced people from various backgrounds. They work for diverse industries including fashion, art, catering, entertainment and technology through concise and contemporary design.

P014–015

NOSIGNER

nosigner.com

NOSIGNER is a social design activist that drives social change towards a more hopeful future. 'NOSIGNER' refers to professionals who recognise the unseen behind the form (the 'SIGN'). They believe design is a tool to form the best relations.

P080–081, 166–167, 236–239

ODDROD

oddrod.es

ODDROD quests for an answer of balance, encompassing the various ways of looking. They seek a kind of enjoyable design that simplifies the complex, expands their gaze, nurtures their experiences, and can be shared with others.

P074–075

Olga Ivanova

behance.net/labazo

Olga Ivanova is a young graphic designer whose main focuses are branding, web design, logo design and typography.

P184–185

P.K.G. Tokyo Inc.

pkg.tokyo

P.K.G. Tokyo Inc. is a brand and design management agency based in Tokyo, Japan.

P062–063

Parámetro Studio

parametrostudio.com

Parámetro Studio provides a fresh contemporary vision that challenges the standards and pushes creativity forwards. They offer art direction, brand identity and consultancy services for clients through visual and conceptual exploration.

P196–197

Plácida

placida.es

Plácida is a graphic design studio focusing on visual identity, packaging and editorial design. They believe in simplicity, harmony and honesty in design to unveil the true essence of any product, concept or brand they work with.

P202–203, 258–259

Plantis Inc.

plantistokyo.com

Founded by designer Keijiro Yamaguchi, Plantis Inc. considers design and manufacturing equally, focusing on packaging, graphic design, stationery design and manufacturing.

P280–282

PMDESIGN

pmdesign.pt

Established in 2008, PMDESIGN is a company specialised in communication design, packaging design, illustration, creation of exhibition environments and graphic system and its application in public spaces.

P274

Polygraphe Studio

polygraphe.ca

Polygraphe Studio is trained at uncovering the smallest detail that will speak to the audience in the most profound, human and unforgettable way. They are sounding boards and hand-holders, who speak from the gut and respect the bottom line.

P170–171

Prompt Design

prompt-design.com

Prompt Design is a branding and packaging team which has won many awards in worldwide competitions. Somchana Kangwarnjit, its founder, has been invited to the international jury panel in many packaging design competitions.

P144–145, 152–153, 162–163, 252–253

Renata Pereira

design-renata.com

Renata Pereira is a multidisciplinary graphic designer who has developed new brands and digital products and partnered with founders and companies to create new products and ventures.

P206–207

RHYTHM INC.

rhythm-inc.jp

RHYTHM INC. is a design studio based in Tokyo, providing branding services, art direction and design works for advertising and promotions.

Rice Studios

thisisrice.com

Rice is a strategic branding and design studio that creates communication and products for commercial and cultural clients. They believe, a powerful and compelling design can transform brands and make positive impacts.

Romina Alfonsi, Carla Estrada, Soha Savant

behance.net/RominaAlfonsi
behance.net/carlaestradagarcia
behance.net/SohaSavant

Romina Alfonsi, Carla Estrada and Soha Savant are designers graduated from Elisava Barcelona School of Design and Engineering with master's degrees.

Ruto design studio

ruto.gr

Oriented by art direction, branding and packaging, Ruto design studio creates beautiful, meaningful and sustainable work for people and brands worldwide through informed, insightful and creative solutions.

SANDU Design

sandudesign.com

Established in 2003, SANDU Design is an integrated design team. They apply design thinking to different fields, provide professional and thoughtful solutions to meet the demands of different projects and communicate their personalities.

Sarah Johnston

sarahleejohnston.com

Sarah Johnston is a designer with over 18 years of professional design experience in visual communication,

product and packaging design, art direction, production, marketing, business and management skills.

SEESAW Inc.

seesaw.jp.net

SEESAW Inc. is a design studio with a strong focus on product and service development. By integrating graphic design, web design, packaging design, promotion and advertising, they offer a solution to develop and grow services and products.

Serious Studio

serious-studio.com

Serious Studio is a branding and design agency for visionaries, building brands with soul and substance. They strive to propel culture, business and people forwards through strategic thinking and intentional design.

Seymourpowell

seymourpowell.com

Seymourpowell is a global strategic design and innovation company with over 30 years' experience of creating award-winning designs, a group of multidisciplinary design researchers, strategists, brand experts, UX and UI designers.

STUDIO L'AMI

studio-lami.com

STUDIO L'AMI specialises in creating great brand identities through seductive print collateral, packaging and environmental design. They believe in meaningful and tactile design experiences, giving people something to touch, hold and feel.

StudioPros

studiopros.work

StudioPros is an award-winning

design agency that provides design services for their clients' brands. They are dedicated to utilising unique, fascinating and precise visual languages to narrate stories and promote the brands for their clients.

studioWMW

studiowmw.com

Working in the global marketplace, studioWMW specialises in brand-building through effective, captivating and conceptual design solutions. They develop trusted relationships with their clients, working to communicating each brand's values.

Tal Nistor

talnistor.com

Tal Nistor is currently a visual communication design student at Shenkar College of Engineering and Design located in Israel. Specialising in branding and digital design, she is a curious, open-minded and eager learner.

The Lab Saigon

thelabsaigon.com

The Lab Saigon is a company of creative professionals in strategy, identity, spatial design, communication and emerging media.

Transwhite Studio

transwhite.cn

Transwhite Studio is a multifaceted studio featuring experimental design and a medium for communication. It is expanding its role to include art exhibitions, social events and other collaborations in various fields.

TSUBAKI KL

tsubakistudio.net

TSUBAKI KL imparts knowledge

to their clients in branding, communications and marketing. It is known for their expertise in graphic design, visual communication and commitment to details with a core focus on brand management.

P082–083

Two Nice Studio

behance.net/twonicedesign

Two Nice is a group of young and hardworking fellows who are eager to craft solutions from a rather playful point of view. They specialise in branding, packaging and editorial design.

P078–079

Unspoken Agreement

unspokenagreement.com

Unspoken Agreement is a collaboration of bold creatives leading the charge in the war against ugliness and visual noise. With a clean and conceptual aesthetic, they strive to create striking design that is both memorable and timeless.

P140–141, 211

Vianka

be.net/viankawu

Vianka is a graphic designer from Indonesia who is always hungry for more as there is always more to learn and more to achieve.

P098–099

Victor Branding Design

victad.com.tw

Victor Branding Design positions itself as 'your design partner' by gathering talents of relevant fields with commitment to online image design, graphic design, packaging design, website design and consultant.

P060–061, 232–233

Victoria Ng

behance.net/ngvictoria

Victoria Ng is a colour fanatic dedicated to creating work that

will bring a spark of joy. She is an imaginative creature, crafty nerd and a curious conceptual thinker who strives to innovate sophisticated designs.

P283

Wing Yang

behance.net/wingyang

Wing Yang is a freelance designer focusing on branding and packaging design. He also enjoys working on animation and illustration.

P020–021, 188–189, 222–223

WWAVE DESIGN

wwavedesign.com

WWAVE DESIGN specialises in brand identities, campaigns and product packaging. They provide rigorous, brand-new designs and consultation services for customers in both commercial and cultural fields.

P096–097, 156–157, 228–229, 264–265

Xiaoxiao Ma

xiaoxiaoma.design

Xiaoxiao Ma is a graphic designer who received her BFA degree in Graphic Design and Digital Media at Academy of Art University in San Francisco.

P308–309

XY Creative

xycreative.cn

XY Creative is a design studio that specialises in branding and packaging design.

P084–089

Y.STUDIO

yuziji.studio

Y.STUDIO is an integrated design agency, specialising in branding, packaging design and editorial design. Their young and flexible thinking provides creative experience for the clients, making the brands more competitive.

P230–231

Yan Zaretsky

munk.design

Yan Zaretsky is a multidisciplinary freelance designer and art director.

P300–301

Yinan Hu

huyinandesign.com

Yinan Hu is a multidisciplinary designer who works on interactive product, funiture, packaging and visual design.

P288–289

Yolanda Sabuz, Sandra Solanas, Sofia Cuba

behance.net/yolandasabuz
behance.net/sandriiita0810
behance.net/sofiacubaree9c

Yolanda Sabuz, Sandra Solanas and Sofia Cuba are designers who were conferred master's degrees in Packaging Design by Elisava Barcelona School of Design and Engineering.

P294–297

Zoltán Visnyai

behance.net/visnyaizoltan

In his works, Zoltán Visnyai gives a visually eye-catching and conceptual response to a given problem. He works and studies in Budapest and is in love with packaging and editorial design.

P212–213

Zoo Studio

zoo.ad

Each of the eight members at Zoo Studio has different yet complementary educational background and professional experience. Their teamwork and individual capacity allow them to work under high standards with a constant spirit of surpassing expectations.

P106–107, 112–115, 168–169, 194–195

ACKNOWL-EDGEMENTS

We would like to express our gratitude to all of the designers and agencies for their generous contribution of images, ideas and concepts. We are also very grateful to many other people whose names do not appear in the credits, but who have made specific contributions and provided support. Without them, the successful compilation of this book would not have been possible. Special thanks to all of the contributors for sharing their innovation and creativity with all of our readers around the world.